Noble Murray Eberhart

Outlines of Economic Entomology

Designed as a Textbook for School and Colleges, and as a Reference-Book

for Farmers and Gardeners

Noble Murray Eberhart

Outlines of Economic Entomology
Designed as a Textbook for School and Colleges, and as a Reference-Book for Farmers and Gardeners

ISBN/EAN: 9783337125356

Printed in Europe, USA, Canada, Australia, Japan

Cover: Foto ©berggeist007 / pixelio.de

More available books at **www.hansebooks.com**

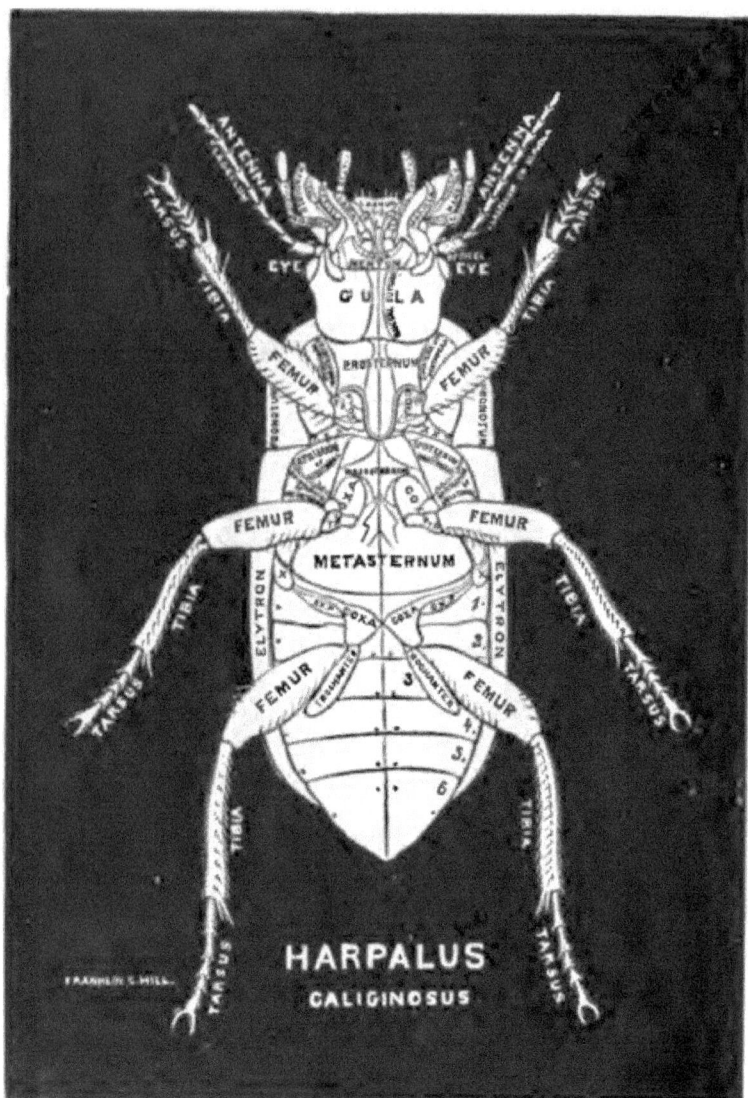

HARPALUS
CALIGINOSUS

EBERHART'S

KEY TO THE

FAMILIES OF INSECTS

ILLUSTRATED.

CHICAGO LAWN, ILL.:
POPULAR PUBLISHING COMPANY,
1888.

PREFACE.

Although there are many Keys to particular orders, and to particular families, there is no Key published, which, in a single volume, starts the young Entomologist on the way to use these specific synopses, and if a student does not know whether his specimen belongs to the **Carabidæ** or to the **Cicindelidæ**, of what use to him is a Key to either of those families?

It is to bridge over this chasm, and to place him in a position to use the particular **Keys**, that this work has been prepared.

N. M. E.

Chicago, September, 1888.

HOW TO USE THE KEY.

If your specimen goes under *A* proceed to *a*, then to *b*, and so on. If it does not go under *A*, try it under *AA* or *AAA*. If it answers to either of these proceed with the next letter under them, as before, and whenever it does not answer to one letter go to the double, triple, etc., of the letter until you find one that it fits.

A KEY TO THE ORDERS OF INSECTS.

A. Mouth fitted for biting, *i.e.*, having jaws and mandibles.

   a. Upper or front wings horny or coriaceous (leathery); under wings membranous.

      b. Upper wings serving to protect the lower, which are folded, first longitudinally (fan-like), and then transversely (doubled under), beneath them.

                                    Coleoptera (*Beetles*).

      bb. Lower wings folded longitudinally, but not transversely.

                    Orthoptera (*Grasshoppers, Crickets, etc*).

   aa. All four wings membranous.

      c. Few-veined wings; end of female abdomen armed with a sting or else with a saw-like ovipositor (the instrument for depositing eggs).

                Hymenoptera (*Bees, Ants, etc*.).

      cc. Wings many-veined; no sting.

            Neuroptera (*Dragon-flies, May-flies, etc*.).

AA. Mouth fitted for sucking

   d. Four wings.

      e. Covered with minute feathers or scales.

           Lepidoptera (*Butterflies and Moths*).

      ee. Not scaled or feathered.

        Hemiptera (*True Bugs, such as Plant-lice, Cicadas, etc*.).

   dd. Two membranous wings.

            Diptera (*Flies, Mosquitoes, etc*.).

   Most authors also add another order, including the little wingless insects known as "bristle-tails" and "spring-tails." The order is called

              Thysanoptera.

A KEY TO THE FAMILIES OF NEUROPTERA.

A. Tarsi (feet), 2 or 3-jointed.

   a. Antennæ ("feelers"), 13-jointed; wings few-veined; species minute.

               Psocidæ.

   aa. Antennæ nearly as long as the body, filiform (thread-like), and many-jointed; wings unequal in size, many-veined.

               Perlidæ.

   aaa. Antennæ short, 5-8-jointed, setiform (bristle-like); wings nearly equal.

           Libellulidæ (*Dragon-flies*).

AA. Tarsi 4 or 5-jointed.

   b. Antennæ quite short.

      c. Antennæ 7-jointed, first two thick, third, long and hair-like.

           Ephemeridæ (*May-flies*).

      cc. Antennæ 20-jointed, somewhat bead-like (sub-moniliform).

           Termitidæ (*White-ants*).

   bb. Antennæ generally long.

      d. Antennæ setaceous (bristle-like); wings large and reticulated (marked like network), anal space (that part of the hind wings nearest the body), plicated (plaited like a fan).

           Sialidæ (*Horned Corydalus, etc*.).

      dd. Antennæ filiform; wings large; posterior (hind) ones have no plicated anal space; ocelli (simple eyes) usually wanting.

        Hemerobidæ (*Aphis-lions, Lace-winged Flies, etc*.).

      ddd. Front of head produced into a slender, deflexed beak, with the mouthparts at the end.

           Panorpidæ.

      dddd. Antennæ filiform; wings longer than the body; mouth imperfectly developed; species sometimes resemble small moths.

           Phryganeidæ (*Caddis-flies, etc*.).

A KEY TO THE FAMILIES OF ORTHOPTERA.

A. Hind legs elongated and adapted to leaping.
    a. Antennæ much longer than body.
        b. Tarsi 3-jointed.

                                            GRYLLIDÆ (*Crickets*).

        bb Tarsi 4-articled.

                                            LOCUSTIDÆ (*Katy-dids*).

    aa. Antennæ shorter than body. Tarsi 3-jointed.

                                          ACRIDIDÆ (*Grasshoppers*).

AA. Legs more or less equal and fitted for walking,—5-jointed tarsi.
    c. Abdomen with strong forceps at the end.

                                          FORFICULIDÆ (*Earwigs*).

    cc. Abdomen without forceps.
        d. Thorax very long, and narrower than the horizontal head; front legs very large and adapted to grasping.

                                    MANTIDÆ (*Praying Mantes*).

        dd. Body very long and slender; head not perceptibly wider than thorax. Legs all about the same size.

                                      PHASMIDÆ (*Walking Sticks*).

        ddd. Head partially concealed under the expanded prothorax (first segment of thorax). Flat body.

                                      BLATTIDÆ (*Cockroaches*).

A KEY TO THE FAMILIES OF HEMIPTERA.

A. Vertical head ; wings membranous or coriaceous throughout.

                                    Sub-order, HOMOPTERA.

AA. Horizontal head ; base of fore or upper wings coriaceous.

                                    Sub-order, HETEROPTERA.

HOMOPTERA.

A. Tarsi 3-jointed.
    a. Antennæ 6 or 7-jointed; ocelli at the corners of a triangle on the back of the head.

                                    CICADIDÆ (*17-year Cicadas*).

    aa. Antennæ 2 or 3-jointed.
        b. Ocelli placed beneath the compound eyes; antennæ 3-jointed.

                                      FULGORIDÆ.

        bb. Ocelli (if present) placed on the forehead; head broad and triangular; antennæ apparently 2-jointed.

                                    CERCOPIDÆ (*Leaf-hoppers*).

AA. Tarsi 2-jointed.
    c. Antennæ 10-jointed, ending in two short bristles.

                                      PSYLLIDÆ (*Leaf-hoppers*).

    cc. Antennæ 7-10-jointed; no bristles.

                                      APHIDÆ (*Plant-lice*).

AAA. Tarsi 1-jointed.

                                      COCCIDÆ (*Bark-lice*).

HETEROPTERA.

A. Tarsi of middle legs, 2-jointed. Aquatic or sub-aquatic species.
   a. Tarsi of front legs, 1-jointed.

NEPIDÆ.

   aa. Hind legs greatly elongated, and strongly fringed with hair.

NOTONECTIDÆ.

   aaa. Eyes on a stem; antennæ small and concealed.

GALGULIDÆ.

   aaaa. Eyes not pedunculated; antennæ small and slender.

HYDROMETRID.N.

AA. Species wholly terrestrial (living on the land).
   b. Beak short and 3-jointed.
     c. Rostrum (beak) fitting in a cavity or trough; body oval and depressed; last joints of antennæ not thinner than the preceding ones.

TINGIDÆ.

     cc. Rostrum naked; head behind narrowed into a more or less elongated neck.

REDUVIIDÆ.

   bb. Beak 4-jointed.
     d. Scutellum large and reaching as far as the middle of the abdomen, often farther.

SCUTELLERIDÆ.

     dd. Scutellum ordinary.
       e. Antennæ on or below a line from eyes to base of beak; ocelli usually absent.
         f. Second joint of antennæ often enlarged at tip; terminal joint thinner than preceding; ocelli always absent.

CAPSIDÆ.

         ff. Last joints of antennæ not thinner than preceding, but not clavated (clubbed). Body usually narrow; ocelli sometimes present.

LYGÆIDÆ.

       ee. Antennæ above line from eyes to base of rostrum; ocelli present.

COREIDÆ.

A KEY TO THE PRINCIPAL FAMILIES OF COLEOPTERA*

A. All of the tarsi usually 5-jointed.
   a. Elytra extending to, or nearly to, the tip of the abdomen.
     b. Apparently 6 palpi, through the outer lobe of the maxillæ being palpiform. Antennæ filiform and simple.
       c. Legs adapted to running; hind trochanters egg-shaped; large and conspicuous.
         d. Head vertical and broader than the thorax; jaws with prominent teeth; generally gaily colored species.

CICINDELIDÆ (Tiger-beetles).

---

*Arranged from Le Baron. The technical names may be understood by referring to frontis piece.

dd. Head not vertical and seldom wider than the thorax; mandibles simple,
or very slightly toothed; color usually black, seldom spotted.

CARABIDÆ (Ground-beetles).

a. Hind legs fringed and fitted for swimming; trochanters not conspicuous.
    e. Hind legs long

DYTISCIDÆ (Diving Beetles).

ee. Hind legs very short.

GYRINIDÆ (Whirligigs).

bb. Outer lobe of maxillæ not palpiform; ant ennæ filiform, but usually serrate.
    g. Prosternum prolonged behind into a point which fits into the
        meso sternum; short legs; very firm body.
        h. Point movable.

ELATERIDÆ (Snapping-beetles).

hh. Prosternal point immovable.

BUPRESTIDÆ.

gg. Prosternal point not prolonged; body soft or moderately firm.
        i. Body moderately hard, with more or less elongated
            legs.

PTINIDÆ.

ii. Body soft.
    j. Antennæ enlarged at tip; palpi clavate

CLERIDÆ - Flower-beetles.

jj. Antenna not enlarged.

LAMPYRIDÆ (Fire-flies).

bbb. Antennæ knobbed or clubbed.
    k. Palpi usually long and exceeding the 6-9
        jointed antennæ in length; aquatic species.

HYDROPHILIDÆ.

kk. Palpi of ordinary length; antennæ usually
    11-jointed; terrestrial species.
    Club of antennæ pectinate (comb-toothed);
    mandibles strongly toothed.

LUCANIDÆ - Stag-beetles.

ll. Club of thin plates; hind legs set back.

SCARABÆIDÆ (Scavenger-beetles.)

lll. Antennæ clavate, but club nearly or quite
    filiform; small species (except Sil-
    phidæ).
    m. Large insects; hind trochanters con-
        spicuous.

SILPHIDÆ - Carrion or Burying-beetles).

mm. Small insects; elytra shorter than
    abdomen.
    n. Head very small and retractile.

HISTERIDÆ.

nn. Head moderate.

NITIDULIDÆ.

mmm. Elytra covering the whole abdomen.

DERMESTIDÆ.

aa Elytra usually covering less than half of the abdomen.

STAPHYLINIDÆ (Rove-beetles).

AA. Hind tarsi 4-jointed; other tarsi 5-jointed.
    o. Head as wide as the thorax and attached to it by a neck body somewhat soft.
        p. Abdomen usually ending in a long point; thorax having a lateral margin.

MORDELLIDÆ.

pp. Abdomen not pointed; side of thorax rounded.

MELOIDÆ (Blister-beetles, etc.).

oo. Head not as wide as thorax and generally partially inserted in it; body firm, oblong.

TENEBRIONIDÆ.

AAA. Tarsi apparently 4-jointed, with all but the last dilated; brush-like underneath.
　　Head more or less prolonged into snout **or beak.**
　　　s. Antennæ elbowed.

　　　　　　　　　　　　　　　　　　　CURCULIONIDÆ *(Weevils).*

　　　　ss. Antennæ not elbowed.

　　　　　　　　　　　　　　　　　　　BRUCHIDÆ (*Pea-weevils, etc.*).

　　rr. Head not having a snout.
　　　　t. Antennæ clubbed or knobbed; **tarsi not dilated.**

　　　　　　　　　　　　　　　　　　　SCOLYTIDÆ.

　　　　　tt. Antennæ generall, filiform or bristle-like (setaceous); tarsi dilated.
　　　　　　　u. Form elongated; antennæ usually long.

　　　　　　　　　　　　　　　　CERAMBYCIDÆ *(Wood-boring Beetles).*

　　　　　　　uu. Form short; antennæ seldom more than half the length of the body.

　　　　　　　　　　　　　　　　CHRYSOMELIDÆ *(Leaf beetles).*

AAAA. Tarsi usually seeming to be 3-jointed; body rounded, or somewhat hemispherical; antennæ very short; elytra usually spotted.

　　　　　　　　　　　　　　　COCCINELLIDÆ *(Lady-birds).*

A KEY TO THE FAMILIES OF LEPIDOPTERA.

A. Antennæ knobbed or clubbed.　　　　　　　BUTTERFLIES.
AA. Antennæ never knobbed or clubbed.　　　Moths.

BUTTERFLIES.

A. With six legs developed for walking.
　　a. Wings erect in repose.
　　　b. Black, white, or yellow colored.

　　　　　　　　PAPILIONIDÆ *(Swallow-tailed Butterflies, etc.).*

　　　　bb. Colors blue or reddish.

　　　　　　　　　　　　　　　　　　LYCÆNIDÆ.

　　aa. Wings usually spread in repose, or, if not, thrown far back.

　　　　　　　　　　　　　HESPERIDÆ *(" Skippers ").*

AA. First pair of legs aborted, leaving only four legs suited to walking.

　　　　　　　　　　　　　　　　　　NYMPHALIDÆ.

MOTHS.

1. Body stout, spindle-shaped; narrow, powerful wings; second or hind pair about half the length of the others.

　　　　　　　　　　　　SPHINGIDÆ *(Hawk-moths).*

2. Wings more or less transparent; small moths found flying in the day time.

　　　　　　　　　　　　　　　　　　ÆGERIDÆ.

3. Head large and free; day-flyers.

　　　　　　　　　　　　　　　　　　ZYGÆNIDÆ.

4. Body thick and heavy; head seemingly sunk in thorax; mouth parts often wanting; antennæ inserted higher up than usual.

　　　　　　　　　　　　　　　　　　BOMBYCIDÆ.

5. Body thick; thorax often crested; antennæ simple or slightly pectinate; wings folded roof-like.

　　　　　　　　　　　　　NOCTUIDÆ *(Owlet-moths).*

6. Broad thin wings, and slender, finely scaled bodies; wings not folded roof like, in repose, but are spread out.

　　　　　　　　　　　　PHALÆNIDÆ *(Geometrids).*

7. Palpi very long, slender and compressed; front pair of legs often tufted.

　　　　　　　　　　　　　　PYRALIDÆ *(Snout-moths).*

8. Palpi very short and beak-like; forewings oblong; antennæ filiform; small species.

　　　　　　　　　　　　TORTRICIDÆ *Leaf rollers.*

9. Wings sickle-shaped (falcate), and edged posteriorly with a heavy fringe; very small or minute species.

　　　　　　　　　　　　　　　　　　TINEIDÆ.

10. Wings fissured and plumed; body long and slender.

　　　　　　　PTEROPHORIDÆ *(Plume-moths).*

A KEY TO THE PRINCIPAL FAMILIES OF HYMENOPTERA.

A. One joint between the coxa and femur.

   a. First joint of hind tarsi more or less cylindrical. not conspicuously wide or hairy.

      b. Front wings flat and not folded.

         c. An erect scale or two knots on the first joint of the abdomen.

                    FORMICIDÆ, (*Ants*).

        cc. Scales or knobs absent.

          d. Prothorax extending to the bases of the front wings.

                 POMPILIDÆ, (*Wasps and hornets*.)

          dd. Prothorax not reaching the bases of the wings.

                    SPHEGIDÆ, (*Mud wasps, etc.*)

      bb. Front wings having a fold lengthwise.

                   VESPIDÆ, (*Wasps and hornets*.)

  aa. First tarsal joint of hind legs somewhat compressed, inner side more or less hairy.

                    APIDÆ (*Bees*).

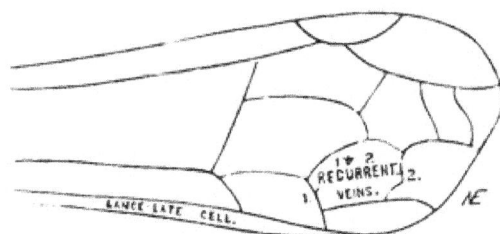

AA. Two joints between coxa and femur.

      e. No lanceolate, and less than three basal or inner cells in front wings.

        f. Recurrent veins absent from front wings.

                 CHALCIDIDÆ, (*Chalcid-flies*).

        ff. One or more recurrent veins present.

                ICHNEUMONIDÆ (*Ichneumon-flies*.)

     ee. Lanceolate cell present.

              TENTHREDINIDÆ (*Saw-flies*.)

* * * * * *

· A KEY ·

Families of Insects

BY

Noble M. Eberhart, B.S., Ph.D.,

Author of "Some Curious Insects," "Outlines of Economic Entomology,"
Etc., Etc., Etc.

CHICAGO LAWN, ILL.:
POPULAR PUBLISHING COMPANY,
1888.

* * * * * *

OUTLINES OF

ECONOMIC ENTOMOLOGY.

DESIGNED AS A TEXT-BOOK FOR

SCHOOLS AND COLLEGES,

AND AS A REFERENCE-BOOK FOR

FARMERS AND GARDENERS.

By NOBLE M. EBERHART, B. S., Ph. D.

*Author of "A Key to the Families of Insects;" "Entomological
Dictionary;" "Comparative Entomology;"
Etc., Etc.*

FULLY ILLUSTRATED WITH ENGRAVINGS MADE EXPRESSLY
FOR THIS WORK.

CHICAGO, ILL.:
A. FLANAGAN, PUBLISHER,
1889.

PREFACE.

So great is the annual loss from the ravages of injurious insects that it seems as though the methods of prevention should be taught in every school.

The aim of the author has been to prepare a concise and practical treatise that may be used in schools where time is limited, and that can be published at a price within the reach of all.

N. M. E.

Chicago Lawn, Ill.,
December 1, 1888.

CONTENTS.

CHAPTER I.

General Features. The branch of the Animal King-dom called *Arthropoda* embraces those animals whose bodies are built upon the plan of rings or segments, which fit together by joints made by the folding in and softening of the outer covering, and giving great flexibility to the body.

Chitine. This outer covering or crust is called chitine, and is the frame-work or skeleton of the insect, answering the same purpose as the bones of the higher animals.

Divisions of the Body. On examining the body of an ant or a wasp, it is at once noticed that the body is divided by a narrowing in of the outlines into three general divisions,—the head, the thorax, and the abdomen.

The Head. In the head are the eyes, and the mastica-tory organs. It also bears the antennæ or "feelers,"—slen-der, hollow and jointed appendages, which are the organs of touch, and, some claim, of hearing also.

The Thorax. The thorax is the seat of the organs of locomotion. To it are attached the wings and legs.

The Abdomen. The abdomen contains the digestive and genital organs. As this work is not intended to set forth in detail the physiology of insects, we will only give a brief outline of their internal system.

The Muscular System. This lies just beneath the chitinous covering or skin of the insect, and according to Newport, it consists of "numerous distinct, isolated, straight fibres, which are not gathered into bundles united by common tendons, or covered by aponeuroses (tendinous sheaths), to form distinct muscles, as in the *Vertebrata*, but remain separate from each other and only in some instances are united at one extremity by tendons."*

The Nervous System. This consists primarily of two longitudinal cords, with a knot of ganglion (nerve centre) for each segment. The position of this is ventral.

The Organs of Nutrition. These are made up of an alimentary canal with its appendages, and are found in various stages of development in different insects, the simplest form being a straight tube.

Circulation. The heart of the insect is a dorsal pulsating tube, terminating in a large artery in the head. The blood of the insect is seldom red,—generally it is colorless,—but sometimes of a yellow tinge.

Respiration. The insect breathes through little tubes or pores called trachea, the terminal openings being called spiracles, of which a row runs along each side of the body, there being normally eleven on each side.

Aquatic insects respire "water mechanically mixed with air," by means of gill-like flattened expansions of the body-wall, called branchiae. Their inner tubes are generally termed bronchial trachea.

The Secretive Organs. Says Packard: "The urinary vessels, or what is equivalent to the kidneys of the higher animals, consist in insects of several long tubes, which empty by one or two secretory ducts into the posterior or

---

* Note. The muscular power of insects is almost incredible. A flea will jump 200 times its own height. Newport mentions an instance where GEOTRUPES STERCORARIUS sustained and escaped from under a pressure of 20 or 30 ounces, the insect itself only weighing about that many grains.

'pyloric' extremity of the stomach. There are also odoriferous glands analogous to the cutaneous glands of vertebrates. The liquid poured out is usually offensive and is used as a means of defense."

Transformations of Insects. We will not follow the development of the embryo, as it is virtually a science in itself, and in this limited work but a general idea can be given. The embryo larva, when it has reached that period that it desires to break from the egg, bursts the shell (which has become somewhat thin at this period), and on emerging begins to feed voraciously. The larva grows rapidly and generally moults, or changes its skin for a number of times. A few days before the assumption of the pupa, or intermediate stage between the worm and the perfect insect, the larva ceases to eat, becomes restless, and either spins a silken cocoon or makes one of earth or chips.

During the semi-pupa state, the skin of the chrysalis grows beneath the nominal covering of the larva. After entering its cocoon it remains in the pupa state a length of time varying with the insect and climate, during which the imago or perfect insect is formed, which finally emerges. The female, after impregnation, immediately provides for the propagation of the species by depositing her eggs in a suitable locality.

CHAPTER II.

Injurious Hymenoptera.

The name *Hymenoptera* is derived from the Greek words, *hymen*, membrane; and *pteron*, wing, (plural, *ptera*). It includes bees, wasps, saw-flies, ants, etc.

They are possessed of greater intelligence, and their transformations are more complete than those of any other order. The larvæ are footless grubs, except in the case of saw-flies, whose young have abdominal legs. The reasoning powers of *Hymenoptera* have been so highly eulogized as to be said to differ from those of man only in *degree*.

THE PEAR-TREE SLUG.

(*Selandria cerasi.* Peck.)

The Pear-Slug hibernates as a pupa, the imagos or perfect insects emerging in May and June. The adult is a bright black fly. If the tree is shaken, the insects usually

Fig. 1. Pear-tree Slug.

fall to the ground and feign death. The saw-flies (to which family the Pear-Slug belongs), are thus named because of the saw-like appendage at the end of the abdomen in most females. With this the leaves of trees are slit, and in these crevices the eggs deposited. Says Saunders: "The female begins to deposit her eggs early in June; they are placed singly within little semicircular incisions through the skin of the leaf, sometimes on the under side and sometimes on the upper. In about a fortnight these eggs hatch. The newly hatched slug is at first white, but soon a slimy matter oozes out of the skin and covers the upper part of the body with an olive-colored sticky coating." A second brood of eggs is deposited late in July. Maturing in about a month they go into the ground and assume the pupa state, in which form they remain during the winter.

Remedies. An ichneumon fly deposits its eggs in those of the Pear-Slug, the grub living in the egg and destroying it. A wash, composed of an ounce of powdered hellebore to each two gallons of water, sprayed on the leaves of the tree is sufficient.

THE IMPORTED CURRANT WORM.
(*Nematus ventricosus.* Klug.)

This is the larva of another saw-fly. The insect generally hibernates as a pupa,—rarely as a grub. The adult insects appear in the beginning of spring. The female is

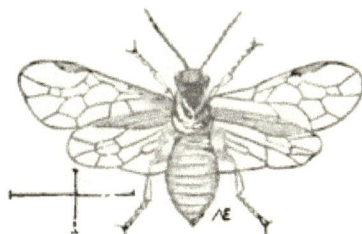

Fig. 2.   Fig. 3.
Imported Currant Worm and Moth (female).

larger than the male, and of a yellow color. The male is spotted with dull yellow. The eggs are placed on the un-

der side of the leaves and on the principal veins. They
hatch in from ten to twelve days. The larvæ eat little
holes in the leaf until nothing but the frame work or skeleton is left. When ready to pupate they form their cocoons
under rubbish; sometimes in the ground, occasionally on
the stems or leaves of the currant bushes. The flies
emerge during the latter part of June or the first of July.
These lay eggs, which soon hatch, and the larvæ generally
change into the pupa state, in which they pass the winter.

Remedies. Parasites prey on the egg and on the
larvæ, notably one found by Prof. Lintner, State Entomologist of New York, which attacks the egg. Dr. Packard
recommends powdered white hellebore sprinkled over the
bushes by means of a muslin bag tied to a stick. Dr Wor-
cester has met with equal success in the use of carbolate of
lime, which was sprinkled over the bushes as soon as the
worm made its appearance. Hand picking is very good.

THE NATIVE CURRANT SAW-FLY.

(Pristiphora grossularia. Walsh.)

We quote from Packard: This saw-fly is a widely dif-

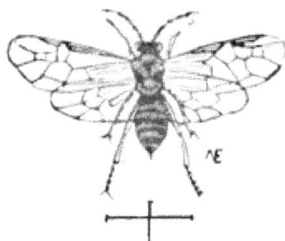

Fig. 4. Native Currant Saw-fly.

fused species in the Northern and Western States, and in-
jures the currant and gooseberry. The female fly is a

shining black, while the head is dull yellow and the legs
are honey-yellow. * * * * Mr. Walsh states that the
larva is a pale grass-green worm, half an inch long, with a
black head, which becomes green after the last moult, but
with a lateral brown stripe meeting with the opposite one
on the top of the head, where it is more or less confluent,
and a central brown black spot on its face. It appears the
last of June and early in July, and a second brood in
August. They spin their cocoons on the bushes on which
they feed, and the fly appears in two or three weeks, the
specimens reared by him flying on the 26th of August.
This worm may at once be distinguished from the imported
currant worm by the absence of the minute black warts
that cover the body of the latter. The same remedies
should be used for this worm as are recommended for the
preceding insect.

CHAPTER III.

INJURIOUS LEPIDOPTERA.

The order *Lepidoptera* (*lepis*, a scale), comprises the butterflies and moths. They are distinguished from other insects in having their wings covered with minute feathers or scales.

The larvæ are seldom footless.

The transformations of *Lepidoptera* are complete.

Moths are distinguished from butterflies, in that the antennæ of the former are pointed at their tip, (occasionally, though, they have small side branches), while the antennæ of butterflies are knobbed or thickened at their ends.

CUT WORMS.

The numerous larvæ passing under this name, belong to the family, *Noctuidæ*, and most of them to the genera *Agrotis* and *Hadena*. They are nocturnal in their habits,

Fig 5.   Cut Worm.

feeding on the roots and tips of herbs. They hibernate, as half-grown larvæ, in oval cavities in the ground. As soon in spring as the frost leaves the ground, they ascend to, or near the surface, and pursue their usual method of feeding; many living entirely on roots, and never coming to the surface; while others travel around during the night, doing great injury by cutting off young herbs near the roots.

When full grown they descend further into the ground than before, where they pupate, emerging in three or four weeks, in the winged state. The eggs are generally deposited on low plants, and when the young larvæ hatch, they go down into the ground and feed upon roots.

Remedies. The Cut-worms have many natural enemies such as the robin, the black-bird, the cat-bird, domestic fowls, some species of ground beetles, (*Carabidæ*); the toad, etc.

Numerous parasites also infest them. Among the artificial remedies, making holes in the ground in the evening with a pointed stick, and going around in the morning and thrusting the stick again into the holes, will destroy any Cut-worms that may have fallen into the holes during the night.

Many will seek shelter at the approach of dawn under leaves and rubbish lying on the ground, these may be found and killed.

Late plowing is good.

The Cut-worms have a great fondness for clover, and little bunches of it, poisoned with arsenic, and placed in the field at night will kill many.

THE CORN WORM.
(*Heliothis armigera.* Hübner.)

This insect is very injurious to the cotton in the South, and there it is called the Boll-worm, because it feeds on the cotton boll. It hibernates as a pupa, a few inches below the surface, in an oval cell lined with silk. In the latitude of Illinois it is two or three brooded, the number increasing with the distance south. The hibernating pupæ become imagos about the time the first shoots of corn appear. They deposit their eggs (which number from fifty to one

Fig. 6.
Corn Worm and Moth.   Fig. 7.

hundred) separately, one on each blade of corn. After hatching, the larvæ eat into the stalk. They get their growth about the time the corn tassels. This brood does so little damage that it generally passes unnoticed.

The eggs of the second brood are laid in the tip of the ear. The larvæ feed on the silk, which appears about the time they hatch.

The best remedy is fall plowing, which exposes the pupæ to the weather.

THE ARMY WORM.

(*Heliophila (Leucania) unipuncta.*)

This is a smooth caterpillar, seldom found later than June or July. It hibernates generally as a larva; occasionally as an imago; rarely as a pupa.

The hibernating larvæ pupate early, about an inch below the surface, and the imagos emerge in March. There are about three yearly broods in the latitude of central Illinois.

The female soon after emerging lays her eggs in grass between the stalk and its surrounding sheath or between the terminal blades while they are yet doubled. Will lay in almost any situation; often in grain, corn-stalks, or hay-stacks.

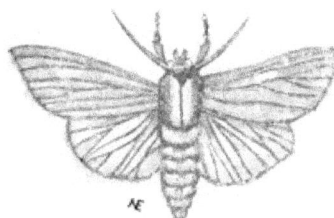

Fig. 8. Fig. 9.

Army Worm and Moth.

The larvæ do not travel in armies unless very numerous, and so often pass unnoticed.

The first brood live as cut-worms, finally going into the ground and undergoing the various stages, the moths appear about the last of June. In five to eight days, these lay.

The third brood emerges in the latter part of August.

In Illinois the second brood does the greatest injury.

They are always more numerous the year following a dry year, but are never destructive and numerous in the same locality for two succeeding years, because of disease and parasites.

Remedies. The tachina fly, the ichneumon fly, the predaceous beetles, and the bobolink destroy many. Fair results in a wheat field, infested by Army Worms, have been obtained by dragging a long rope over the top of the stalks, jarring the worms to the ground. This repeated twice a day in small fields and where the worms are not too thick would prove advantageous, but in larger fields it is a question whether the results would equal the time and labor expended.

A very good method is the plowing of a furrow around the fields.

The worms will collect in this ditch and a log dragged along it with a rope will crush large numbers of them. Poisoning plants around the edges of a field with a mixture of Paris green and water is useful where the worms are not too numerous.

THE FALL ARMY WORM
(Laphygma frugiperda.)

is covered with stiff erect hairs and appears only in the fall, therefore it may be easily distinguished from the foregoing.

THE PEACH-TREE BORER.
(Sannina (Egeria) exitiosa.)

Hibernating in the pupa state, the moth appears in May and June. The eggs which are of a beautiful yellow-brown color are deposited singly on the trunks of the peach and cherry, near the roots, and held in position by a gummy secretion. They are about one fiftieth of an inch long, and a little more than half as wide.

Fig. 10.   Peach-tree Borer (female).

The larva as soon as it hatches seeks a crevice and works down under the bark toward the roots.

It is, according to Saunders, "a naked soft cylindrical grub, of a pale whitish yellow color, with a reddish, horny looking head, and black jaws." The presence of the larva is readily detected by the exudation of gum.

Remedies. The larvæ may be sought for and killed, directly. Hot water is recommended by many. In using it, the dirt should be scraped away from the roots, and the water poured on hot enough that it will not cool before reaching the grubs.

A wash of carbolic acid and soap suds is also useful.

By far the best method, however, is to mound up the ground about the trees to the height of a foot or so, which will prevent the female from laying her eggs.

THE STALK-BORER.

(*Gortyna nitela.* Guenee.)

This larva may be easily distinguished because the stripes which on most cut-worms run the whole length of the body, are on the Stalk-Borer interrupted for four segments.

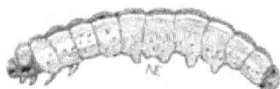

Fig. 11.   Fig. 12.

Stalk-Borer and Moth.

It hibernates as a moth, which comes forth early in spring and lays its eggs on blue grass and young grain. The young larvæ often do great injury by eating off the terminal blades of young oats.

Pupating in the ground in August, the imagos emerge by the latter part of August or the first of September.

The best remedy is keeping down such weeds and grass
as the eggs are liable to be deposited on.

The larvæ are preyed upon by a parasite.

THE CANKER WORM.

(*Anisopteryx vernata.*)

The Spring Canker Worm hibernates as a pupa, a short
distance below the surface of the earth.

Some emerge from October on, if the winter is a mild
one The remainder generally do so about the middle of

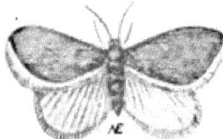

Fig. 13.   Fig. 14.

Male and Female Canker Worm Moths.

March, the females appearing first.   These females are
wingless.   They crawl up the tree and deposit their eggs in
irregular masses of fifty to one hundred between branches,
under scales of bark, or in any other sheltered situation.

These eggs are of a broad oval form.   They hatch in a
few days, about the time the apple trees are leafing out, and
immediately attack the leaves, puncturing them with small
holes.   They soon strip the trees of most of their leaves,
and will kill them in two or three seasons.

The color of the larva is a dark olive-green or brown,
much resembling the general color of the tree.   They fasten
themselves by their two pairs of posterior or pro-legs to a
twig, and hold the body away from the tree so that they resemble a short spine or branch.   This is a protective device.

When alarmed they drop by a slender silken thread a

few inches, so that they are out of the reach of their enemies. They mature in two or three weeks, when they
descend and pupate.

There is but a single brood. They sometimes attack
plum, cherry and elm trees.

The best and most effective remedy is to spray the larvæ
with arsenical poison.

THE STRAWBERRY LEAF ROLLER.

(Phoxopteris comptana.)

This insect generally hibernates in the pupa state, rolled
up in the strawberry leaf; but sometimes as an imago.

Those that have wintered as pupæ emerge in April and
May, and deposit their eggs in May and June; the larvæ

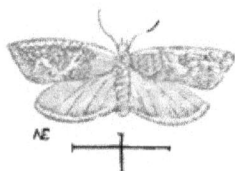

Fig. 15. Moth of Strawberry Leaf Roller.

getting their growth in July. The second brood matures
late in September.

The only effectual remedy is to mow the strawberry field
close, after the fruit has been picked, and after letting the
grass, etc., become dry, burn it.

FOREST TENT CATERPILLAR.

(Clisiocampa sylvatica, Harris.)

The Forest Tent Caterpillar hibernates in the egg state,
the larvæ often emerging before the leaves on the tree are
out; but are able to fast from a week to ten days, and so

suffer no injury from their early appearance.   There is but
a single yearly brood.   When there are large numbers of
larvæ they swarm and defoliate acres of shrubbery.   They

Fig. 16.

Fig. 17.   Fig. 18.

Forest Tent Caterpillar, Moth and Egg mass.

attain their growth in June, pupate, emerge the same year,
and lay their eggs in vertical belts around the twigs. These
belts are covered with a mucous which is sometimes so
thick that the eggs cannot be seen.

COMMON TENT CATERPILLAR

(C. Americana,)

of our orchards does not essentially differ in its habits.   Its
egg masses may be distinguished from the fact that they

Fig. 19.   Fig. 20.

Moth and Egg-mass of the Common Tent Caterpillar.

taper at the ends, instead of ending abruptly and verti-cally.

They are called Tent Caterpillars because they spin a web, living in a community under it and going out twice a day to feed.

Remedies. The tents may be destroyed or the larvæ killed as they crawl over the trunks of the trees. The twigs bearing the egg masses may be cut off and destroyed. The best method, however, is spraying the trees with arsenical poison, which will not only destroy the larvæ but also many other injurious insects infesting the tree.

THE TOMATO WORM.

(Macrosila quinquemaculata.)

This insect is also frequently called the potato worm. The pupa (in which form it hibernates), is readily recog-nized by the case in which the tongue develops being bent

Fig. 21.   Fig. 22.

Tomato Worm and Pupa. (About one-fourth natural size.)

around so that it resembles the handle of a pitcher. The larva is a large green caterpillar with oblique whitish stripes on the sides and a horn on the anal extremity. The imago emerges in June and July.

The larvæ are so large that hand-picking is a good rem-edy. The moths may be caught with a net.

THE CODLING MOTH.

(Carpocapsa pomonella, Linn.).

The Codling Moth is easily distinguished from other moths by a large egg-shaped spot, brown in color, edged

with copper, and situated on the hinder margin of each
fore-wing. It generally hibernates as a pupa, emerging in
the spring about the time the petals of the apple-blossoms fall.
The female lays her eggs in the calyx or eve of the

Fig. 21. Larva of Codling Moth.

forming apple. These eggs hatch in about a week and the
grub eats into the core. The larvæ become full-grown in
three or four weeks. About this time the prematurely
ripened fruit falls to the ground.

Sometimes the worm escapes before and sometimes not
until after the fruit has fallen. Those leaving before crawl
down the trunk, or lower themselves by a silken thread,
which they have the power of spinning.

The first and last segments of the body are at first black
but become brown as the grub matures; the other segments
each have six or eight spots on them, from which arise little
hairs.

The larva pupates in a cocoon placed in a crevice of
the bark or in some other sheltered place. There are two
yearly broods.

Remedies. Ichneumon flies destroy some. The fallen
fruit should be gathered or the hogs allowed to devour it,
thus destroying many larvæ. The best method, however,
is to put bands around the trunks of the trees. The larvæ
will pupate in these and may be gathered and destroyed.
These bands should be examined every ten days or less
from the last of May to the last of August.

CABBAGE BUTTERFLIES.

In the list of injurious insects, Cabbage Butterflies occupy a prominent place, because their field of operation is so extensive, and the means of exterminating them, as yet, so imperfect.

THE EUROPEAN CABBAGE BUTTER-FLY
(Pieris rapæ.)

was first noticed in this country in the year 1857 by a Mr. Bowles of Quebec. Not long after, it spread into New England and New York, and a few years later it was plentiful all over the country.

It passes the winter in the pupa state, the perfect insect emerging early in the spring. There are about five broods

Fig. 24.

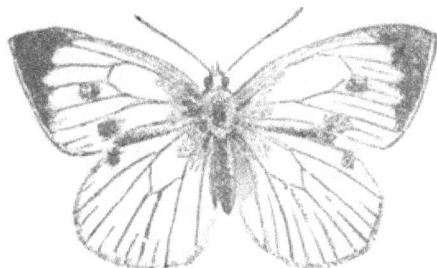

Fig. 25.

European Cabbage Butterflies (male and female).

during the year, in the latitude of Illinois. This number increases as we go farther south, and vice versa.

The female insect is distinguished from the male in that it has two spots on its wings, while the male has only one.

These eggs are usually laid on the upper side of a cabbage leaf, and are not collected in a mass in one place,

Fig. 26.   Larva of European Cabbage Butterfly.

but are scattered over the surface of the leaf. When about to pupate, the larvæ seek shelter under boards lying in the field, or under the copings of walls and fences.

Remedies. By far the greater number of Cabbage Butterflies are destroyed by a parasite (*Pteromalus puparum*), which lays its eggs on the pupa, and the little maggots hatching out eat their way into the body of the insect, an operation attended with much pain. They devour the fatty portions, thus preventing the pupa from transforming into the perfect state.

By placing boards in the cabbage field, the pupæ, which will soon be found on the under side of these, may be collected and placed in a box covered with a screen, to allow the parasites to hatch out and escape, while at the same time the cabbage insect cannot.

Of late years another natural remedy is rivalling the parasite mentioned above, for its efficacy in disposing of the Cabbage Butterfly.

This is a contagious disease which is prevalent among the larvæ, and destroys them in a short time. Instances are known where a whole field has been entirely cleared of larvæ in twenty-four hours.

The symptoms of the disease are that shortly after midsummer the larva has an ashy appearance, and later becomes greenish milky.

After death, which occurs in a few hours, the body dries or shrivels up and on being touched crumbles to pieces.

Professor S. A. Forbes, the State Entomologist of Illinois, has been making a series of experiments, by trying to breed the bacteria of this disease in distilled water, and then to clear an infested field by communicating the bacteria to some of the larvæ. The success of the experiment has not yet become established, but it is hoped that it soon will be.

The larvæ may be destroyed by sprinkling them with a mixture of pyrethrum and water, which has the advantage of killing the worms, and at the same time it is perfectly harmless in its effects on the human race, so that no evil results come from sprinkling it on the cabbages. A child with a net can do a great deal of good by capturing the butterflies.

THE SOUTHERN CABBAGE BUTTERFLY

(*Pieris protodice.*)

is a native of this country, and does not differ essentially in the habits from the European or imported species, but it is far less injurious.

Fig. 27. Southern Cabbage Butterfly (female). (One-fourth natural size.)

Some gardeners have found sawdust impregnated with carbolic acid, an efficient remedy.

The tachina fly is another parasite similar in its opera-

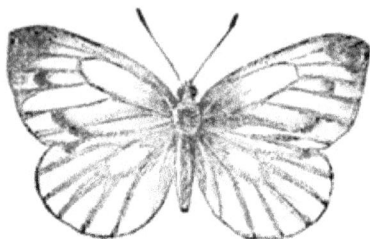

Fig. 8.   Southern Cabbage Butterfly (male).

tions to the chalcid fly (*Pteromalus puparum*), mentioned above.

THE CABBAGE PLUSIA.

(*Plusia brassica*   Riley.)

"In the months of August and September," says Professor Riley, "the larvæ may be found quite abundant on this plant (cabbage), gnawing large, irregular holes in the leaves. It is a pale green translucent worm, marked longitudinally with still paler, more opaque lines, and like all the known larvæ of the family to which it belongs, it has but two pairs of abdominal pro-legs, the two anterior segments, which are usually furnished with such legs in ordinary caterpillars, not having the slightest trace of any, consequently, they have to loop the body in marching, as represented in the figure, and are true "Span-worms." Their bodies are very soft and tender, and as they live exposed on the outside of the plants, and often rest motionless, with the body arched, for hours at a time, they are espied and devoured by many of their enemies, such as birds, toads; etc.   They are also subject to the attacks of at least two parasites, and die very often from disease, especially in wet weather, so that they are never likely to increase quite as badly as the butterflies just now described.

Fig. 29.   Fig. 30.

Cabbage Plusia and Larva.

"When full-grown, this worm weaves a very thin, loose white cocoon, sometimes between the leaves of the plant on which it fed, but more often in some more sheltered situation, and changes to a chrysalis, which varies from a pale yellowish green to brown, and has a considerable protuberance at the end of the wing and leg cases, caused by the long proboscis of the inclosed moth being bent back at that point. This chrysalis is soft, the skin being very thin, and it is furnished at the extremity with an obtuse roughened projection which emits two converging points, and several short curled bristles, by the aid of which it is enabled to cling to its cocoon.

"The moth is of a dark smoky-gray, inclining to brown, variegated with light grayish-brown, and marked in the middle of each front wing with a small oval spot and a somewhat U-shaped silvery white mark, as in the figure. The male is easily distinguished from the female by a large tuft of golden hairs, covering a few black ones, which springs from each side of his abdomen towards the tip.

"The suggestions given for destroying the larvæ of the cabbage butterflies, apply equally well to those of the Cab-

bage Plusia, and drenching with a cresylic wash will be
found even more effectual, as the worms drop to the ground
with the slightest jar."

THE MELON WORM.
(*Phacellura hyalinatilis.* Linn.)

This insect hibernates as a pupa rolled up in the leaves
of some plant or tree. Says Willet: " The Melon Worms
are of a light, yellowish-green color, nearly translucent, have
a few scattered hairs, and when mature, are about an inch
and a quarter in length.   They 'web up' in the leaves of
the melon, or of any plant growing near, which has flexible
leaves, forming a slender brown chrysalis, three-quarters of
an inch in length.   Hundreds of these pupæ were found

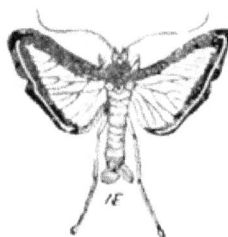

Fig. 31.   Fig. 32.

Melon Worm and Moth.

rolled up in the leaves of the tomato and sweet potato.   In
passing through one of the patches referred to, numbers of
small, beautiful moths rose from the grass and weeds.
Their wings when extended measured an inch across, and
were of an iridescent pearly whiteness, except a narrow
black border.   Their legs and bodies presented the same
glistening whiteness, and the abdomen terminated in a
curious movable tuft of white appendages, like feathers, of

a pretty buff color, tipped with white and black. These moths proved to be the mature Melon Worms which had emerged from the chrysalids referred to."

Remedies. Plant early, and pick off the first brood of worms by hand. An ichneumon-fly (*Pimpla conquisitor*) and a species of tachina-fly are two parasites which prey upon the Melon Worm.

THE PALMER WORM.

(*Ypsolophus pometellus*. Harris.)

During the beginning of summer or the latter part of spring, greenish ochre-colored larvæ may be noticed feeding on the leaves of the apple and cherry trees.

The following account of this insect, which is commonly

Fig. 33. Moth of the Palmer Worm.

called the Palmer Worm, is taken from Saunder's Insects Injurious to Fruits:

"It lives in societies, making its home in a mass of half-eaten and browned leaves, drawn together by silken threads, from which it drops when the tree or branch is jarred, suspended in the air by a thread of silk. The larva is of a pale yellowish-green color, with a dusky or a blackish stripe along each side, edged above by a narrow whitish stripe; there is also a dusky line along the middle of the back. Its head is shining yellow, and the top of the next segment is of the same color; on each ring there are several small black dots, from each of which arises a fine yellow hair. While young the caterpillars eat only the green pulpy tissue of the leaves, leaving the net-work of veins entire; later on they consume the whole of the leaf except its coarser veins. They also frequently gnaw holes or irregular cavities in the young apples. These larvæ feed on the leaves of the cherry as well as those of the apple.

"When full grown they are about half an inch long. They then change to chrysalids within the mass of eaten leaves occupied by the larvæ, and ordinarily spin a slight cocoon in a fold of a leaf, but when they are very abundant the foliage is so entirely consumed that they have to look for shelter elsewhere. Their chrysalids are then often found under dry leaves on the surface of the ground, in crevices in the bark of the tree, and in other suitable hiding-places. The chrysalis is about a quarter of an inch long; at first it is of a tawny yellow color, which gradually changes to a darker hue. In ten or twelve days the perfect insect is produced.

"The moth is of an ash-gray color. The fore wings are sprinkled with black atoms, and have four black dots near the middle, and six or seven smaller ones along the hinder margin. The hind wings are dusky above and beneath, with a glossy azure-blue reflection, blackish veins, and long, dusky fringes. The antennæ are alternately striped with black and white. Sometimes the fore wings are of a tawny yellow, in other specimens they are tinged with purplish red, and in some the dots are faint or entirely wanting. They rest with their long, narrow wings folded together and laid flat upon their backs."

Remedies. "Showering the trees with whale-oil soap and water has been recommended, but the use of Paris green and water would prove more effectual; the water would dislodge many of the larvæ, and the remainder would be destroyed by eating the poisoned leaves."

THE AMERICAN SILK WORM.
(Telea polyphemus. Linn.)

Also commonly known in the adult state as the Poly-phemus Moth.

The full-grown caterpillar is a very large worm often approximating four inches in length. It feeds on the leaves of the plum tree, and has been reared somewhat extensively for its silk.

Remedies. Many natural enemies prey upon it while in the larva state. It is never very injurious, its economic interest lying principally in the fact of its being raised for silk, but if it should prove destructive to the plum trees in any locality it may be readily gotten rid of by hand picking.

THE GREEN GRAPE VINE SPHINX.
(Darapsa myron. Cramer.)

We quote from Saunders, having had no opportunity of personally observing the habits of this insect: "The larva is one of the most common and destructive of the leaf eating insects injurious to the grape. The first brood of the perfect or winged insects appears from the middle to the end of May, when the female deposits her eggs on the under side of the leaves, generally placing them singly, but sometimes in groups of two or three. The eggs are nearly round, about one-twentieth of an inch long, a little less in width, smooth, and of a pale yellowish-green color, changing to reddish before hatching. The young caterpillar

comes out of the egg in five or six days, when it makes its
first meal on a part of the empty egg shell, and then attacks the softer portions of the grape-vine leaves. When
first hatched, it is one-fifth of an inch long of a pale yellowish-green color, with a large head and having a long
black horn near its posterior extremity, half as long as its
body. As it increases in size, the horn becomes relatively
shorter and changes in color; the markings of the larva
also vary considerably at each moult. When full grown it
is about two inches long, with a rather small head of a pale
green color, dotted with yellow, and with a pale-yellow stripe
down each side; the body is green, of a slightly deeper
shade than the head, and covered with small yellow dots
or granulations; along the sides of the body these granulations are so arranged as to form a series of seven
oblique stripes, extending backwards, and margined behind
with a darker green. A white lateral stripe with a dark-
green margin extends from just behind the head to the
horn near the other extremity. Along the back are a series
of seven spots, varying in color from reddish to bluish
green, granulated with black in front, and sometimes yellow
behind and at the tip. This larva has the power of draw-
ing the head and next two segments within the fourth and
fifth, causing these latter to appear much distended; the
feet are red, the pro-legs pale green. Some specimens
especially among those of the later brood, will be found
exhibiting remarkable variations in color; instead of green
they assume a delicate reddish-pink hue, with markings of
darker shades of red and brown, which so alter their appearance that they might at first be readily taken for a different species; a careful comparison, however, will show the
same arrangements of dots and spots as in the normal

form. When full grown, the larva descends from the vine and draws a few leaves closely together, binding them with silken threads, usually about or near the base of the vine on which it has fed, and within this rude structure changes to a chrysalis of a pale-brown color, dotted and streaked with a darker shade, and with a row of oval dark brown spots along each side.

The moths from this first brood of larvæ usually appear during the latter part of July, when they deposit eggs for a second brood, which mature late in September, pass the winter in the pupa state, and emerge as moths in the following May.

The wings of this insect, when fully expanded, measure about two and a half inches across, their form being long and narrow.

The fore wings are of a dark olive-green color, crossed by bands and streaks of greenish gray, and shaded on the outer margin with the same hue. The hind wings are dull red, with a patch of greenish gray next the body, shading gradually into the surrounding color. On the under side the red appears on the fore wings, the hinder pair being greenish gray. The antennæ are dull white above, rosy below, head and shoulder covers deep olive-green, the rest of the body of a paler shade of green; underneath the body is dull gray.

This moth rests quietly during the day, taking wing at dusk, when it is extremely active; its flight is very swift and strong, and as it darts suddenly from flower to flower, rapidly vibrating its wings, remaining poised in the air over the objects of its search, while the long, slender tongue is inserted and the sweets extracted, it reminds one strongly of a humming bird.

"The caterpillars are very destructive to the foliage of the vine, being capable of consuming an enormous quantity of food; one or two of them, when nearly full grown, will almost strip a small vine of its foliage in the course of two or three days. In some districts they are said to nip off the stalks of the half grown clusters of grapes so that they fall unripe to the ground.

Remedies. "The readiest and most effectual method of disposing of these pests is to pick them off the vines and kill them. They are easily found by the denuded canes which mark their course, or where the foliage is dense they may be tracked by their large brown castings, which strew the ground under their places of resort. Nature has provided a very efficient check to their undue increase, in a small parasitic fly, a species of Ichneumon, the female of which punctures the skin of the caterpillar and deposits her eggs underneath, where they soon hatch into young larvæ, which feed upon the fatty portions of their victim, avoiding the vital organs. By the time the Sphinx Caterpillar has become full grown, these parasitic larvæ have matured, and eating their way through the skin of their host, they construct their tiny snow white cocoons on its body, from which, in about a week, the friendly fly escapes by pushing open a nicely fitting lid at one end of its structure. No larvæ thus infested ever reaches maturity; it invariably shrivels up and dies.

THE AMERICAN PROCRIS.

(*Procris Americana.* Harris.)

Most of the insects hibernate in the pupa state; a few as imagos.

Those that winter as pupæ emerge during June and

deposit their eggs in patches of 20 or more on the under side of the leaves of the grape. The larvæ soon hatch and feed in flocks on the back of the leaves.

"While young, the little caterpillars eat only the soft

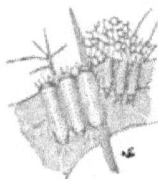

Fig. 34. Larvæ of American Procris.

tissues of the leaves, leaving the fine net-work of veins untouched, but as they grow older they devour all but the larger veins." (Saunders.) They mature in August, and pupate in a crevice in the bark. In a fortnight the moths emerge, and a second brood of larvæ soon follows. The majority of these remain through the winter as chrysalids.

Remedies. Spray the vines thoroughly with Paris green and water (one-half teaspoonful of Paris green to a gallon of water). A parasitic fly destroys the larvæ.

THE IMPORTED CURRANT BORER.

(*Egeria tipuliformis*. Linn.)

The imago is a pretty, wasp-like moth, measuring about three-quarters of an inch across the expanded wings. The body is bluish black, with three yellow bands across the

Fig. 35.

Fig. 36.

Imported Currant Borer and Moth.

abdomen. It appears about the middle of June. The female deposits her eggs singly close to the buds.

They burrow into the stem and bore up and down, feeding on the pith. (Saunders).

They pupate in the stem, having first eaten a hole nearly through to the outer air, so that when the moth is about to appear it can easily burst through and escape.

Remedy. Cut and burn all hollow stems found in the fall or spring.

THE GOOSEBERRY FRUIT WORM.

Dakruma convolutella. Hübner.

This insect hibernates as a pupa, the moth appearing the last of April or the first of May. The female lays her egg on the young gooseberries, the larva burrowing into the fruit. Only a single hole is made in a berry.

When alarmed the worm backs out quickly and drops

Fig. 37. Gooseberry Fruit Worm.

down a few inches by a silken thread which it spins. Sometimes it drops entirely down to the ground. It pupates in a little brown cocoon amid the rubbish on the surface of the ground.

Remedies. Hand picking and the destroying of infested berries.

Sprinkling air-slaked lime on the bushes in early spring is useful in preventing the female from laying. Renew if the rain washes it off.

CHAPTER IV.

INJURIOUS DIPTERA.

The order *Diptera*, ("two-winged") includes the mosquito, the gnat and the common house-fly, the Hessian fly. etc. Also the *Syrphus* and *Tachina* flies which are useful because they destroy many injurious insects.

The larvæ of *Diptera* are called maggots.

The distinguishing feature of this order is that the second pair of wings are not developed as in other orders, but are rudimentary, serving as "balancers."

THE HESSIAN FLY.

(Cecidomyia destructor. Say.)

"This insect is double-brooded, as the flies appear both in spring and in autumn. At each of these periods the fly lays twenty or thirty eggs in the leaf of the young wheat plant.

"In about four days in warm weather they hatch, and the pale-red larvæ crawl down the leaf, working their way in between it and the main stalk; passing downward till they come to a joint, just above which they remain, a little below the surface of the ground, with the head toward the root of the plant. Here they imbibe the sap by suction alone, and by the simple pressure of their bodies they become imbedded in the side of the stem. Two or three larvæ thus imbedded serve to weaken the plant and cause it to wither and die.

"The larvæ become full grown in five or six weeks, then measuring about three-twentieths of an inch in length.

About the first of December their skin hardens, becoming
brown; and then turns to a bright chestnut color. This is
the so-called flax-seed state or puparium.   In two or three

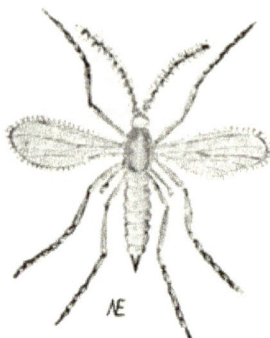

Fig 38.   Fig. 39.

Hessian Fly and Maggot. (Very highly magnified.)

weeks the 'larva,' (or, more truly speaking, the semi-pupa),
becomes detached from the old case. In this puparium
some of the larvæ remain through the winter. Toward the
end of April or the beginning of May, the pupa becomes
fully formed, and in the middle of May in New England,
comes forth from the brown puparium, 'wrapped in a thin
white skin,' according to Herrick, 'which it soon breaks
and is then at liberty.'   The flies appear just as the wheat
is coming up; they lay their eggs for a period of three
weeks, and then entirely disappear. The maggots hatched
from these eggs take the flax-seed form in June and July,
and are thus found in the harvest time, most of them remaining on the stubble. Most of the flies appear in autumn," (From Packard's Injurious Insects of the West,
p. 696.)

   Remedies. There are a number of parasites of the

Hessian fly, which have done a great deal of late years to check its ravages.

The predaceous beetles, swallows and martens destroy many.

Changing or rotating crops is advantageous.

THE BLACK ONION FLY.

(*Ortalis flexa.* Wied.)

The fly is approximately half an inch in length, each wing having three whitish, oblique, crescent-shaped bands or stripes

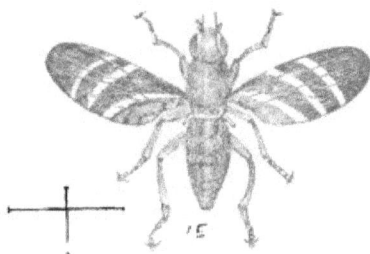

Fig. 40. Black Onion Fly. (Lines show real size.)

There are two yearly broods.

The maggots of the first brood may be found during the month of June. They remain from twelve to fourteen days in the pupa state.

The imagos are rather slow of flight and do not fly any great distance.

Remedies. The only remedies that have as yet been tried with any success are the careful removal of all infested onions, and the use of the kerosene emulsion (see Chapter VIII).

The application of salt, in the proportion of three or four bushels to the acre, has proved useful.

THE IMPORTED ONION FLY.

(Anthomyia ceparum. Bouche.)

The eggs of this species are deposited on the bases of the leaves during May and June. The larvæ appear soon and proceed to eat their way down to the base of the young bulb.

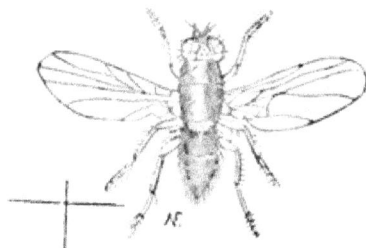

Fig. 11. Imported Onion Fly.

In about fourteen days they pupate in the ground, and a couple of weeks later the second brood of flies appear, which generally lay their eggs on the bulb itself.

Remedies.   Same as for the black onion fly.

The sickly onions are readily known by their turning yellow.

THE RADISH FLY.

(Anthomyia radicum. Bouche.)

"Soon after the early radishes come up," says Dr. Packard, "the roots are attacked by small white maggots, and when the plants grow in old soil, the maggots are especially destructive.   The larvæ appear in the spring as soon as the radishes are partly grown."

"When full-grown they change in the ground to reddish-brown pupæ, similar to those of the onion and cabbage maggots.   The insect remains in this state two or three weeks, when the fly hatches and crawls up out of the ground, with its wings crumpled up, and climbing up the side of a clod or any perpendicular surface which it finds, these members expand and assume their proper form before they become dried and firm." (Dr. Fitch's Eleventh Report.)

Remedies. Destroy all infested roots.

Salt and lime sprinkled on the plants will be found useful. Planting early avoids the evil to a great extent, as does also the rotation of crops.

CHAPTER V.

INJURIOUS COLEOPTERA.

Coleoptera (*coleos*, a sheath) are so named because the front wings are usually horny and opaque and cover over or shield the back or membranous pair, which are folded longitudinally and transversely beneath them.   These wing covers are called *elytra*, (singular, *elytron*.)   The members of the order are called beetles.   The common or popular term for the larva is "grub" or "borer."

THE CORN ROOT WORM.

Diabrotica longicornis.

The beetle may be found in corn fields in August and September feeding on fallen pollen and thistles and other composite plants.

About the middle of September the females deposit their eggs in little clusters in the ground, at the bases of the hills

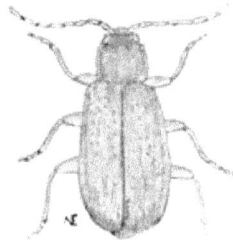

Fig. 42   Adult Form of the Corn Root Worm.   (Very highly enlarged.)

of corn.   These eggs are about one-fortieth of an inch long, resembling minute hen's eggs.   A microscopic examination will show that they are covered with little six-sided pits.

The eggs are entirely unprotected, and yet they weather the winter, hatching out in the spring about the time the corn begins to grow. The full grown larva is nearly half an inch long. It burrows into the roots of the corn, mining lengthwise, and causing them to decay unless the season is very wet.

The worm pupates in August, in an oval cell in the ground, and the beetle hatching out, commences to feed on the pollen of the corn, frequently devouring the silk, and if they are not too hard, the grains of corn also.

As the Corn Root Worm always stays in the same locality, and does not move about much, there is a very simple and effective remedy against it, viz., the changing of crops, which will soon start out the Root Worms.

STRAWBERRY ROOT WORMS.

There are three genera of *Chrysomelidæ*, known as Strawberry Root Worms, each occurring at different times, as follows:

Colaspis..............April—June.
Paria ....June—August.
Scelodonta............August—June.
(Active from August - October.)

They much resemble the larva known as the crown borer, but the latter is footless, and so they are easily distinguished from it.

The eggs are laid in the ground at different periods of the year, according to the species, the larvæ feeding on the roots of the strawberry leaves.

The genus *Scelodonta* feed only on the strawberry.

Paria also feed on the juniper, and *Colaspis* on the grape.

Colaspis hibernates in the egg state; *Paria* as an imago, and *Scelodonta* as a mature larva. They all may be destroyed with Paris green.

WIRE WORMS.

The Wire Worms belong to the family *Elateridæ*, and to the genera *Melanotus, Corymbites* and *Agriotes*. They are hard, smooth and cylindrical, with acute senses, and possessed of great activity.

Fig. 43. Wire Worm.

They live three years in the larva state, when they pupate in the earth, emerging from June to September.

Remedies. Attract and destroy them by slices of poisoned potato.

They do not injure corn until the second year of planting on the grass land, and letting the land lie fallow for a time is a good remedy.

THE APPLE-TREE BORER.

(*Saperda candida*. Fabr.)

This insect is also called the round-headed borer. The eggs are laid in the bark at the bottom of the tree, during May and June. The larvæ bore upward into the wood, where they remain for two or three years, when pupating in a little cocoon some eight or ten inches from their starting place, they emerge during midsummer.

Remedies. Digging out the larvæ is recommended.

Soft soap and soda, mixed with water to the consistency of paint, and applied once in June, and once in July is effective.

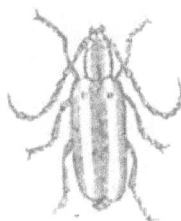

Fig. 44.   Fig. 45.

Round-headed Apple-tree Borer and Beetle.

THE FLAT HEADED APPLE-TREE BORER.

(Chrysobothris femorata. Fabr.)

The Flat Headed Borer lives one year, hibernating in a hole in the wood of the tree. It pupates from April to June,— in Illinois about May,—emerging in June and July. The

Fig. 46.   Fig. 47.

Flat-headed Apple-tree Borer and Beetle.

adult is a flat beetle, which deposits its eggs either singly or in patches on the bark or under scales.

The larvæ hatch in a few days and burrow in the sapwood. A few will soon destroy the tree.

The remedies are the same as for the round-headed borer. The larvæ are found near the bottom of the trunk.

THE PLUM CURCULIO.

(Conotrachelus nenuphar. Herbst.)

The Plum Weevil hibernates in the adult state. The beetle is a short thick one, with a rough-surface, and much

Fig. 48 Plum Curculio. (Greatly enlarged. Showing also the crescentic cut in the fruit.)

resembles a dried bud. It is distinguished from the apple curculio by having two humps on the back. The female

Fig. 49. Larva of Plum Curculio.

makes a hole in the fruit with her snout, in which she lays her eggs, and then makes a crescentic cut around the place.

The best remedy is to spread sheets under the tree, and hit the trunk, jarring off the beetles which may be collected and burned.

THE APPLE CURCULIO.

(*Anthonomus quadrigibbus.* Say.)

The Apple Curculio has four humps and makes a round puncture in the fruit, in which the eggs are deposited.

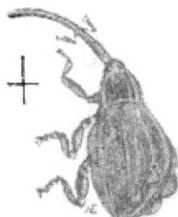

Fig. 50. Apple Curculio.

The larvæ go down to the core. They pupate in about a month in the fruit, and a fortnight or so later the perfect insect appears.

The only remedy known as yet to be at all useful is to jar the infested apples off the tree and feed them to the swine.

THE PLUM GOUGER.

(*Coccotorus scutellaris.* LeC.)

The Plum Gouger is somewhat similar in habits to the plum curculio. Its footless larvæ bore into the seed, where

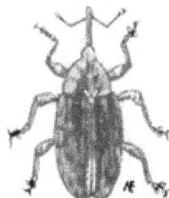

Fig. 51. Plum Gouger.

they live. The same remedies apply that are used for the plum curculio.

THE PEACH CURCULIO.

(*Ithycerus novehoracensis.* Forster.)

Also commonly known as the New York Weevil. It is the largest species of snout-beetle occurring in this country.

Fig. 52.   Fig. 53.

Larva and Imago of the Peach Curculio or New York Weevil.

The beetles appear in May and June, doing considerable injury to the buds and twigs of the peach-tree, although frequently found in the apple, plum, pear and cherry. The female makes a hole in the twig under the bark in which she deposits an egg. The larvæ are footless.

Remedies. The same as for the plum curculio.

THE POTATO BEETLE.

(*Doryphora 10-lineata*, Say.)

Also commonly called the Potato-bug and the Colorado Potato Beetle, the last name being the correct one.

It is too well known to require any description, but a few points may be stated which most people are not familiar with.

They pass the winter in the perfect or beetle state, remaining dormant in the ground, and appearing early in spring. The females deposit their eggs on the under side of the leaves and the orange-colored eggs hatch in about a week into little grubs. These begin feeding on the leaves, and maturing in two or three weeks descend to the ground and pupate under rubbish or in the earth.

They remain from ten to twelve days in this state, when they emerge and the process is repeated,—the number of yearly broods varying, sometimes being as many as four or five, and at other times only two or three.

A closely allied beetle (*Doryphora juncta*, Germar), is often mistaken for the real one, but this latter feeds on various species of Solanum, (the genus which includes the ground-cherry, horse-nettle, etc.,) and never attacks the potato.

Remedies. Paris green or some other arsenical poison is the most effectual. One pound of it should be mixed with twenty of pulverized plaster, or with common flour, and dusted on the leaves in the early morning, the dew holding it there.

It may also be applied to advantage just after a shower. A duster made of a tin box, with a perforated bottom, and a handle four or five feet long, is recommended.

Care should be taken not to inhale any of the green, as it is a deadly poison.

London purple may be substituted for the green and will be found as effective as well as considerably cheaper.

THE PEA WEEVIL.
(*Bruchus pisi*, Linn.)

"The Pea Weevil," says Treat, "is easily distinguished from all other species of the genus with which we are troubled, by its larger size, and by having on the tip of the abdomen * * * two dark oval spots, which cause the remaining white portion to look something like the letter

T.   It is about 0.18–0.20 inch long and its general color is
a rusty black, with more or less white on the wing covers,
and   *   *   *   on the hinder part of the thorax, near the
scutel.   *   *   *   It is supposed to be an indigenous North
American insect and was first noticed   *   *   *   around
Philadelphia, from whence it has spread over most of the
state where the pea is cultivated.   The female deposits her
eggs on the *outside* of the pod.   It is a very general re-
mark that peas are "stung by the bug" and the impression
prevails   *   *   *   that the female punctures and deposits her eggs in the pea in which the larva is to be nour-

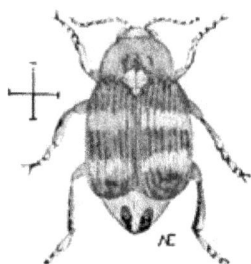

Fig. 54.   Fig. 55.
Pea Weevil and Larva.

ished."   The beetles appear about the time the peas blos-
som and the yellow eggs are laid on any part whatever of
the *surface* of the pod, being held there by fluid which is
rather viscid, and on drying is white and glistening, quot-
ing again from Treat: "The newly hatched larva is of a
deep yellow color, with a black head, and it makes a direct
cut through the pod into the nearest pea.   The hole soon
filling up in the pod, and leaving but a mere speck, not as
large as a pinhole, in the pea.   The larva feeds and grows
apace, and generally avoids the germ of the future sprout,
perhaps because it is distasteful so that most of the buggy

peas will germinate as readily as those that have been untouched. When full grown the larva * * * eats a circular hole on one side of the pea, and leaves only the thin hull as a covering. It then retires and lines its cell with a thin and smooth layer of paste, pushing aside and entirely excluding all excrement, and in this cell it assumes the pupa state, and the beetle when ready to issue has only to eat its way through the thin piece of hull which the larva had left covering the hole. It has been proved that the beetle would die if it had not, during its larval life, prepared this passage way, for Earnest Menault asserts that the beetle dies when the hole is pasted over with a piece of paper even thinner than the hull itself."

Remedies. Take care that no buggy peas are planted. Put them in water; the sound ones sink and the buggy ones float on top and may be readily skimmed off. In localities where few of your neighbors raise peas, or where they consent to do the same as you, if you plant no peas at all for a year or two the bugs will be effectually gotten rid of, or at least lessened so that they will do comparatively little damage.

THE ASPARAGUS BEETLE.

(*Crioceris asparagi.* Linn.)

Hibernating in the adult state the females deposit their first eggs in May. The larvæ hatch in about a week.

The eggs are blackish and the larvæ a sombre ash color. They feed on the bark on the young shoots of asparagus. In the latter part of June they pupate in slight cocoons under rubbish or in the earth. The second brood of larvæ emerge usually between August the 10th and 20th and the beetles mature in September.

Remedies.   "A small shining black parasitic fly " destroys large numbers.   Destroying in early spring all young
shoots or seedlings, in fact all plants but the more mature

Fig. 56.   Asparagus Beetle.

marketable ones, is effectual, as the female must perforce
deposit her eggs on the latter, and as these are cut and sold
every few days the eggs are not allowed to hatch in the
field.

THE STRIPED CUCUMBER BEETLE.

(Diabrotica vittata.   Fabr.)

This insect is universally distributed, and wherever found
is looked upon by the cucumber raiser as his greatest enemy.

The adult beetles appear early in the spring and at once
proceed to their destructive occupation.

Fig. 57.   Striped Cucumber Beetle.

They are said to frequently devour the terminal shoot of
the sprouting seed thus effectually destroying the plant.

The larvæ, which are hatched later on, are whitish grubs,
about half an inch long.   Becoming full grown in about a
month after hatching.

They pupate in cells in the ground. There are two or three yearly broods.

Remedies. The cheapest and most effectual remedy is to cover the plants with boxes, open at the bottom and covered with netting.

Sprinkling the plants early in the day with a mixture of two parts of Paris green and eight parts of flour is recommended.

THE GRAPE-VINE FLEA-BEETLE.
(Graptodera chalybea. Illg.)

Hibernates in the adult state. Comes forth early in the spring and feeds on the buds as soon as they commence swelling.

In three or four weeks it deposits its eggs in little clusters on the under sides of the leaves. The eggs are yellowish in color, and " in a few days produce colonies of small, dark-brown larvæ, which feed on the upper side of the leaves, riddling them; and when numerous they devour the whole

Fig. 58.   Fig. 59.
Grape-vine Flea-beetle and Larva.

leaf except the larger veins, and sometimes entirely strip the vines of foliage."

In about a month the larvæ mature, "when it is a little more than three-tenths of an inch long, usually of a light brown color, sometimes dark, and occasionally paler and yellowish. The head is black, and there are six or eight shining black dots on each of the other segments of the body, from each dot arising a single brownish hair. The under surface is paler than the upper, its feet, six in number, are black, and there is a fleshy, orange-colored pro-leg on the terminal segment."

"When mature, the larvæ leave the vines and descend to the ground, where they burrow into the earth and form small, smooth, oval cells, within which they change to dark-yellowish chrysalids."--(Saunders.)

The beetles appear in a fortnight or so and feed upon the leaves. They are possessed of highly developed thighs, which enables them to jump to some distance, and on account of which they receive their name.

Remedies. Spray the plants thoroughly in the spring with Paris green and water (a teaspoonful to a gallon).

The absence of rubbish about the vines will remove the shelter which the beetles seek and thus be beneficial.

Sprinkle air-slaked lime around the vines in the fall.

CHAPTER VI.

INJURIOUS HEMIPTERA.

Hemiptera (" half-winged " insects) have a portion of the upper or front wings thick and coriaceous (leathery). The members are called "true bugs," and among them are the chinch-bug, squash-bug, bed-bug and plant-louse.

The larvæ are like the perfect insect except that they have no wings.

THE CHINCH BUG

(*Blissus leucopterus.* Say.)

The Chinch Bug is by far the most formidable enemy with which the raiser of corn has to contend. It would appear that nothing can be devised to control the ravages

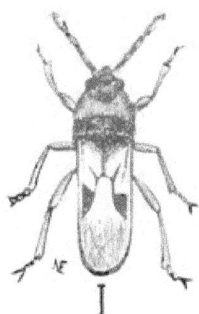
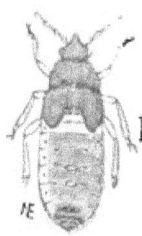

Fig. 60.　　　　　　　　　Fig. 61.

Chinch Bug and Pupa.

of the insect, but as long as a possibility remains, the economic entomologist will seek for the panacea.

As a brief sketch of the life history of the insect, we quote as follows from S. A. Forbes, State Entomologist of Illinois, who has made a careful study of its life and habits for some years. " The eggs are usually laid early in spring, on the roots or lower part of the stem of grain in the field, and to these the young are confined for a time after they hatch.

As they get larger and more numerous, they come out of the ground and gather on the stalks of the wheat or oats, remaining there until the ripening of the grain compels them to seek food elsewhere. At this time they are commonly just beginning to acquire wings, but they migrate to the corn field on foot, as a very general rule, gathering for the first few days on the outer rows of the field. As soon as the larger part of the brood acquire wings, however, they begin to scatter through the field, laying their eggs on the corn, where the second brood live in the cornfields until cold weather approaches, when they scatter everywhere for shelter under which to pass the winter. In the spring they emerge and deposit their eggs in the grain fields as already described."

Remedies. Wet weather has proved very destructive in its effects upon Chinch Bugs; no actual experiments having been made, however, but this is shown by the testimony of past years. Frequently plowing and harrowing a narrow strip of land bordering the field has offered an obstruction to those Chinch Bugs which enter the field in masses and on foot. Another method is to place boards on edge around the field and keep their upper edges daubed with coal-tar. As the bugs of the first brood remain at first on the borders they may be reached here with insecticides, and their destruction prevents the second brood from developing.

Professor Forbes found "that a simple mechanical mixture of water and three per cent. of kerosene " was deadly to bugs of all ages, nor did it injure the corn, provided the kerosene was well emulsed.

The corn should be sprayed with this fluid. He found the cost of this mixture to be about four mills per gallon. With proper appliances the cost ought not to exceed five dollars an acre. And "if by treating a strip at the outer edge of a corn field,—the few rows nearest a ripening field of wheat for example,--the whole field could be protected against the savage ravages of the bugs, it would certainly pay the farmer well to undertake this task."

The greatest practical results, however, will probably be obtained through the natural enemies of the Chinch Bug. It was found that lady-bugs and predaceous ground-beetles destroy a considerable number.

But by far the most deadly enemy of the Chinch Bug is a species of bacteria. This infests the stomach and other internal organs, and much resembles that found by Pasteur in the silk worm. Professor Forbes' method is to cultivate this silk worm virus for the destruction of various insect pests.

HARLEQUIN CABBAGE BUG.

(*Murgantia (Strachia) histrionica.* Hahn.)

This insect derives its name from its gay colors and harlequin-like manners.

The eggs are about one-twentieth of an inch in length, and very beautiful little fellows too. They are laid in two parallel rows of some half a dozen each. Says Riley: " When first deposited they are green in color, but soon become white, with black markings. Their resemblance to miniature white barrels with black hoops is very marked, and the resemblance is heightened by a small black spot in the proper position for a bung hole. The sides of the eggs which are applied to each other are almost entirely black.

In oviposition the female moves her ovipositor in a zigzag manner from one row to the other. The young larva in hatching cuts out the head of the barrel with its beak with the utmost neatness and precision."

Fig. 62. Harlequin Cabbage Bug.

This insect has a great preference for such plants as the cabbage and the turnip; but has no aversion to mustard and radishes. ·

Remedies. Hot water is very good, as is also the method of entrapping them under leaves and rubbish where they have sought shelter.

Burning weeds and rubbish and care and cleanliness in cultivation are useful.

The kerosene emulsion might be tried.

THE TARNISHED PLANT BUG.

(Lygus lineolaris. Beauv)

This destructive insect is plentiful all over the country. It attacks a variety of plants, doing great injury.

Hibernating in the mature state, they deposit their eggs

in early spring, and both old and young bugs may be
found together during most of the summer. The young
ones do not differ from their parents, except in being en-

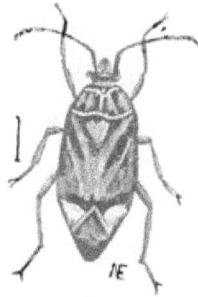

Fig. 63. Tarnished Plant Bug.

tirely green in color, and without wings. There are probably two broods during the year.

"This bug is a very variable species, the males being
generally much darker than the females. The more common color of the dried cabinet specimens is a dirty yellow,
variegated * * * with black and dark brown; and one
of the most characteristic marks is a yellow V, sometimes
looking more like a Y, or indicated by three simple dots on
the scutel, (the little triangular piece on the middle of the
back, behind the thorax.) The color of the living specimens
* * * frequently inclines to olive-green. The thorax,
which is finely punctured, is always finely bordered and
divided down the middle with yellow lines, very frequently
obsolete behind. The thighs always have two dark bands or
rings near their tips."—(Riley.)

Remedies. Pyrethrum is effective against this insect,
as is also the kerosene emulsion, provided it contains not less
than five per cent. of the kerosene.

THE COTTONY MAPLE SCALE.

(*Pulvinaria innumerabilis.* Rathvon.)

" The young lice hatch in spring or early summer, walk about actively as soon as born, and settle along the ribs of the leaves (very rarely on the young twigs). They then insert their beaks and begin to pump up sap and to increase in size, a thin layer of waxy secretion immediately beginning to cover the dorsum. In a little more than three weeks they have increased to double their size at birth, and undergo their first moult, shedding the skin, it is supposed, in small fragments. After this first moult, the waxy secretion increases in abundance and a differentiation between the sexes is observable. The males grow more slender and soon cease to increase in size, covering themselves with a thick coating of whitish wax. The pupa then begins to form within the larval skin, the appendages gradually taking shape, the head separating from the thorax, the mouth parts being replaced by a pair of ventral eyes. A pair of long wax filaments is excreted from near the anus and these continue to grow during the life of the insect. It is the protrusion of these filaments from beneath the waxy scale which indicates the approaching exclusion of the male. The posterior end of the scale is in this manner raised up, and the perfect insect backs out with its wings held close to the sides of its body.

" Meanwhile the female larvæ * * * grow larger and also broader across the posterior portion but remain flat. * * *   Just before the appearance of the adult males they undergo another moult, and change in color from a uniform pale yellow to a somewhat deeper yellow with deep red markings."*

---

* C. V. Riley, Report of U. S. Entomologist, 1884.

Remedies. Spray the trees with the kerosene emulsion, late in May or early in June. The bark louse has a number of natural enemies, such as the predaceous beetles, the lady-bug, a species of harvest mite, and two true parasites.

THE SQUASH BUG.

(*Anasa (Coreus) tristis.* Degeer.)

The females deposit their yellowish brown eggs in June (in the latitude of Illinois), cementing them to the under sides of the squash leaves. The young bugs moult their skins a number of times and at last attain the adult state without passing through the dormant pupal stage. The perfect insects are a rust-colored brown above or rather

Fig. 61. Squash Bug. (Somewhat enlarged.)

yellow, so covered by tiny black dots, that it appears to be a rusty black. The color of the under side of the body is yellow. They are readily known by the odor they emit which resembles that of the banana.

They live upon the juices of the leaves which they suck up through their beaks causing the leaves to wither and die.

Remedies. Hand picking of the bugs, and the destruction of the eggs which are to be found on the under side of the leaves.

THE APPLE APHIS.

(Aphis mali. Fabr.)

"During the winter," says Saunders, "There may often be found in crevices and cracks of the bark of the twigs of the apple tree, and also about the base of the buds, a number of very minute, oval, shining black eggs. These are the eggs of the apple tree aphis, *Aphis mali.* They are deposited in the autumn, and when first laid are of a light yellow or green color, but gradually become darker, and finally black.

As soon as the buds begin to expand in the spring, these eggs hatch into tiny lice, which locate themselves upon the swelling buds and the small, tender leaves, and inserting their beaks feed on the juices. All the lice thus hatched at this period of the year are females, and reach maturity in ten or twelve days, when they commence to give birth to living young, producing about two daily for two or three weeks, after which the older ones die. The young locate about the parents as closely as they can stow themselves, and they also mature and become mothers in ten or twelve days, and are as prolific as their predecessors. They thus increase so rapidly that as fast as new leaves expand, colonies are ready to occupy them. As the season advances, some of the females acquire wings, and, dispersing, found new colonies on other trees. When cold weather approaches, males as well as females are produced, and the season closes with the deposit of a stock of eggs for the continuance of the species for another year. When newly born the Apple Aphis is almost white, but soon becomes of a pale, dull greenish-yellow. The mature females are generally without wings; their bodies are oval in form, less than one-tenth of an inch long, of a pale yellowish-green color, often striped with deeper green. The eyes are black, honey tubes green, and there is a short, tail-like appendage of a black color." The winged females and the males are very similar in color. The head, thorax and antennæ are black, with the neck usually green. The abdomen is short and thick, of an oval form and bright green color, with a row of black dots along each side; the wings are transparent, with dark brown veins.

Most of the insects belonging to this family (Aphidæ) are provided with two little tubes or knobs, which project one on each side, from the hinder part of their bodies; these are called honey tubes, or nectaries, and from them is secreted in considerable quantities a sweet fluid. This fluid falling upon the leaves and evaporating gives them a shiny appearance, as if coated with varnish, and for the purpose of feeding upon this sweet deposit, which is known as honey-dew, different species of ants and flies are found visiting them. Ants also visit the colonies of aphides and stroke the insects with their antennæ to induce them to part with some of the sweet liquid, which is greedily sipped up. This fluid is said to serve as food for a day or two to the newly-born young.

The leaves of trees infested by these insects become distorted and twisted backwards, often with their tips pressing against the twig from which they grow, and they thus form a covering for the aphides, protecting them from rain. An infested tree may be distinguished at some distance by this bending back of the leaves and young twigs. It is stated that the scab on the fruit of the apple tree often owes its origin to the punctures of these plant lice. This species, which was originally imported from Europe, is now

found in apple orchards all over the Northern United States
and Canada.

Remedies. Lady-bugs destroy many. Syringe trees
in spring when buds are bursting with weak lye, strong
soap-suds, or tobacco-water.

THE GRAPE PHYLLOXERA.

(*Phylloxera vastatrix.* Planchon.)

"The insect* presents itself under several different
forms, all of which belong to two types. One of these is the
Leaf-Gall Type (gallicola), and the other is found upon the
roots of the vine (radicicola).

"First, as to the Leaf-Gall Type (*Gallicola*). The gall
or excrescence produced by this is a fleshy swelling of the
under side of the leaf, more or less wrinkled and hairy,

Fig. 65. Female Gall Louse. (Very highly magnified.)

with a corresponding depression of the upper side, the
margin of the cup being fuzzy, and drawn together so as
to form a fringed mouth. It is usually cup-shaped, but
sometimes greatly elongated or purse-shaped. Soon after
the first vine leaves that put out in the spring have fully
expanded, a few scattering galls may be found, mostly on
the lower leaves, nearest the ground. These vernal galls

*This article is condensed by Mrs. Treat from Prof. Riley and here copied
by us.

are usually large (of the size of an ordinary pea), and the normal green is often blushed with rose where exposed to the light of the sun. On carefully opening one of them, we shall find the mother-louse diligently at work surrounding herself with pale yellow eggs of an elongate oval form scarcely one hundredth of an inch long, and not quite half as thick. She is about four hundredths of an inch long, generally spherical in shape, of a dull orange color, and looks not unlike an immature seed of the common purslane. At times by the elongation of the abdomen, she is more or less perfectly pear-shaped. Her members are all dusky, and so short, compared to her swollen body, that she appears very clumsy, and undoubtedly would be outside of her gall, which she never has occasion to quit, and which serves her alike as dwelling house and coffin. More carefully examined, her skin is seen to be shagreened or minutely granulated and furnished with rows of minute hairs. The eggs begin to hatch, when six or eight days old, into active little oval, six-footed beings, which differ from their mother in their brighter yellow color and more perfect legs and antennæ, the tarsi being furnished with long, pliant hairs, terminating in a more or less distinct globule. In hatching, the egg splits longitudinally from the anterior end, and the young louse, whose pale-yellow is in strong contrast with the more dusky color of the egg-shell, escapes in the course of two minutes. Issuing from the mouth of the gall, these young lice scatter over the vine, most of them finding their way to the tender terminal leaves, where they settle in the downy bed which these leaves afford, and commence pumping up and appropriating the sap. The tongue sheath is blunt and heavy, but the tongue proper—consisting of three brown, elastic, and wiry filaments, which, united, make so fine a thread as scarcely to be visible with the strongest microscope—is sharp, and easily run into a leaf Its puncture causes a curious change in the tissues of the leaf, the growth being so stimulated that the under side bulges and thickens, while the down on the upper side increases in a circle around the louse, and finally hides and covers it as it recedes more and more within the deepening cavity. Sometimes the lice are so crowded that two occupy the same gall. If, from the premature death of the louse, or other cause, the gall becomes abortive before being completed, then the circle of thickened down or fuzz enlarges with the expansion of the leaf, and remains to tell the tale of the futile effort, otherwise in a few days the gall is formed, and the inheld louse, which, while eating its way into house and home, is also growing apace, begins a parthenogenetic maternity by the deposition of fertile eggs, as her immediate parent had done before. She increases in bulk with pregnancy, and one egg follows another in quick succession until the gall is crowded. The mother dies and shrivels, and the young, as they hatch, issue and found new galls. This process continues during the summer until the fifth or sixth generation. Every egg brings forth a fertile female, which soon becomes wonderfully prolific. The number of eggs found in a single gall averages about two hundred; yet it will sometimes reach as many as five hundred. Even supposing there are but five generations during the year, and taking the lowest of the above figures, he immense prolificacy of the species becomes manifest. As summer advances, they frequently become prodigiously multiplied, completely covering the leaves with their galls. The lice also settle on the tendrils, leaf-stalks and tender branches, where they also form knots and rounded excrescences much resembling those made on the roots. In such a case the vine loses its leaves prematurely. Usually, however, the natural enemies of the louse seriously reduce its numbers by the time the vine ceases its growth in the fall, and the few remaining lice, finding no more succulent and suitable leaves, seek the roots. Thus, by the end of September the galls are mostly deserted, and those which are left are almost always infested with mildew, and eventually turn brown and decay. On the roots, the young lice attach themselves singly or in little groups, and thus hibernate. The male gall louse has never been seen, and there is every reason to believe that he has no existence. Nor does the female ever acquire wings. It is but a transient state, not at all essential to the perpetuation of the species, and does, compared with the other type, but trifling damage. As already indicated, the autumnal individuals of *Gallicola* descend to the roots, and there hibernate. There is every reason to believe also that, throughout the summer, some of the young lice hatched in the galls are passing on the roots; as considering their size, they are great travelers, and show a strong disposition to reach the earth with ease and safety. At all events, we know from experiments, that the young *Gallicola*, if confined to vines on which they do not normally form galls, will, in the middle of summer, make themselves perfectly at home on the roots.

THE ROOT INHABITING TYPE.

(*Radiciola.*)

We have seen that, in all probability, gallicola exists only in the wingless, shagreened, non tubercled, fecund female form. *Radiciola*, however, presents itself in two principal forms. The newly hatched larvæ of this type are

undistinguishable, in all essential characters, from those
hatched in the galls; but in due time they shed the smooth
larval skin, and acquire raised warts or tubercles which at
once distinguish them from gallicola. In the development
from this point the two forms are separable with sufficient
ease: one of a more dingy greenish yellow, with more
swollen fore-body, and more tapering abdomen: the other
of a brighter yellow, with the lateral outline more perfectly
oval, and with the abdomen more truncated at tip. The
first or mother form is the analogue of gallicola, as it never
acquires wings, and is occupied, from adolescence till death,

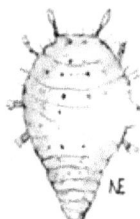

Fig. 66.   Somewhat Mature Larva of the Root-inhabiting Type.   (Very highly
magnified.)

with the laying of eggs, which are less numerous and somewhat larger than those found in the galls. We have counted
in the spring as many as two hundred and sixty-five eggs in
a cluster, and all evidently from one mother, who was yet
very plump, and still occupied in laying.   As a rule, however, they are less numerous.   With pregnancy this form
becomes quite tumid and more less pyriform, and is content
to remain with scarcely any motion in the more secluded
parts of the roots, such as creases, sutures, and depressions,
which the knots afford.   The skin is distinctly shagreened
as in *Gallicola.*   The warts, though usually quite visible

with a good lens, are at other times more or less obsolete, especially on the abdomen.

The second or more oval form is destined to become winged. Its tubercles, when once acquired, are always conspicuous; it is more active than the other, and its eyes increase rather than diminish in complexity with age. From the time it is one-third grown, the little dusky wing pads may be discovered, though less conspicuous than in the pupa state, which is soon after assumed. The pupæ are still more active, and, after feeding a short time, they make their way to the light of day, crawl over the ground and over the vines, and finally shed their skin and assume the winged state. In this last moult the tubercled skin splits on the back, and is soon worked off; the body in the winged insect having neither tubercles nor granulations. These winged insects are most abundant in August and September, but may be found as early as the first of July, and until the vines cease growing in the fall. The majority of them are females, with the abdomen large and more or less elongate. From two to five eggs may invariably be found in the abdomen of these, and are easily seen when the insect is held between the light, or mounted in balsam or glycerine. A certain proportion have an entirely different shaped and smaller body, the abdomen being short, contracted, and terminating in a fleshy and dusky protuberance; the limbs stouter, and the wings proportionately larger and stouter. This form has been looked upon as the male. As fall advances the winged individuals become more and more scarce, and as winter sets in, only eggs, newly hatched larvæ, and a few wingless, egg-bearing mothers are seen. These last die and disappear during the winter, which is mostly passed in the larva state, with here

and there a few eggs. The larvæ thus hibernating become
dingy, with the body and limbs more shagreened and the
claws less perfect than when first hatched; and, of thousands
examined, all bear the same appearance, and all are furnished with strong suckers. As soon as the ground thaws
and the sap starts in the spring, these young lice work off
their winter coat, and growing apace commence to deposit
their eggs. Since, in 1870, the absolute identity of these
two types was proved by showing that the gall-lice become
root-lice. The fact has been repeatedly substantiated by
different observers. (In 1873 galls were obtained on the
leaves of a Clinton vine from the root-inhabiting type, thus
establishing the identity of the two types.)

THE MORE MANIFEST AND EXTERNAL EFFECTS OF PHYLLOXERA DISEASE.

The result which follows the puncture of the root louse
is an abnormal swelling, differing in form according to the
particular part and texture of the root. These swellings,
which are generally commenced at the tip of the rootless,
eventually rot, and the lice forsake them and betake them
selves to fresh ones—the living tissue being necessary to
the existence of this as of all plant lice. The decay affects
the parts adjacent to the swellings, and on the more fibrous
roots cuts off the supply of sap to all parts beyond. As
these last decompose, the lice congregate on the larger
ones, unti. at last the root system literally wastes
away.

Remedies. Thus far, the only practicable method of
combating the insect when established upon the root, is by
drowning it by irrigating the soil. In Europe the method
largely adopted is to graft their vines upon varieties, the

roots of which are Phylloxera proof; for this purpose American varieties have been sent to Europe in immense numbers, as cuttings and as rooted plants. An enterprising grape growing firm has even established nurseries in Europe for the production of vines that resist the Phylloxera.

CHAPTER VII.

INJURIOUS ORTHOPTERA.

Orthoptera ("straight-winged" insects,) include the locusts, grasshoppers, crickets, cockroaches, etc.

The upper wings are more or less leathery, and protect the lower ones, which are folded fan-like beneath them.

As in *Hemiptera*, the larvæ differ from the adults only in the absence of wings.

LOCUSTS.

(Acridida.)

The abdomen of the female locust is armed with an ovipositor (the organ used in depositing eggs), consisting of four horny valves, two curving upward and two downwar   When ready to lay her eggs, she makes a hole in the ground with this ovipositor, in which they are deposited one at a time, placed obliquely and in regular order, so as to form an oval mass.

Fig. 67.   P. femur-rubrum.

The eggs are covered with a white mucus, which ultimately hardens and holds them together.

The hole above the cluster is then closed, the soil being mixed with this same mucus, which, hardening, prevents the accession of moisture.

The eggs in the mass are placed in four rows, that part toward the surface which will allow the newly hatched insects to emerge head-foremost.

The masses are generally placed in hard and compact earth in preference to that which is loose or sandy.

When the locusts are plentiful, the females may even be found boring into the hard soil of a well traveled street.

The young locusts resemble the adults in every respect except that they have no wings.

In a few hours after hatching they begin to feed on whatever appropriate food they find near them.

Fig. 68. P. Spretus.

They develop rapidly, moulting or casting their skin repeatedly, until they attain the adult state, the wings appearing at the second or third moult.

The locusts devour all varieties of vegetation, and great destruction is attendant on their appearance.

The common red-legged species (*Pezotettix (Caloptenus) femur-rubrum*, De Geer), prefers to feed upon grasses in open areas, while the Rocky Mountain Locust (*P. spretus*, Thomas), a closely allied species, differing principally in having longer wings, feeds upon any plant that comes in its way.

Remedy. The most effectual remedy has been demonstrated to be the kerosene emulsion (see next chapter).

NOTE. The following paragraphs on the locusts, from

the Ninth Report of the State Entomologist of Illinois, by
Dr. Thomas, may be of interest to many readers:

"CLIMATIC INFLUENCE."

"Dampness is undoubtedly the most potent natural
agent in keeping them in check.

Although they may have hatched out in excessive numbers, yet if a rainy season follows soon afterwards, they
will be destroyed to a very large extent, and the invigorated
vegetation will bid defiance to the feeble attacks of those
that remain alive. Like other insects their breathing ap-
paratus consists of tubes that permeate the body, connect-
ing with opening or breathing pores along the sides of the
body, one on each side of a segment.

The moisture taken in by inspiration in all probability
produces disease, or at least in some way prevents the free
passage of the air and thus lessens the vitality.

Excessive changes during the winter also appear to have
a tendency to destroy the vitality of the eggs. That those
of the red-legged and other allied species, which are somewhat boreal in their habits, can withstand a greater degree
of cold, is undoubtedly true, but they are certainly affected
by sudden and considerable changes.

CHAPTER VIII.

KEROSENE EMULSIONS.

This remedy has become so popular of late years that it is certainly deserving of a special chapter.

It stands at the head of the Economic Entomologist's list of insecticides.

The methods of emulsifying kerosene were first made public in 1880, and since that time they have come into universal use.

"It cannot be too strongly impressed upon all who use kerosene as an insecticide," says Riley, "that it can be considered a safe remedy only when properly emulsified."

The great point to be looked after is that there is sufficient agitation to make a permanent emulsion.

The following formula of Riley's is that which Mr. Hubbard found so satisfactory in destroying the scale-insects infesting the orange:—

Kerosene.................... 2 gallons = 67 per cent.
Common or whale oil soap..½ pound } = 33 per cent.
Water...................... 1 gallon }

"Heat the solution of soap and add it boiling hot to the kerosene. Churn the mixture by means of a force-pump and a spray-nozzle, for five or ten minutes. The emulsion, if perfect, forms a cream, which thickens on cooling, and should adhere without oiliness to the surface of glass. Dilute before using, one part of the emulsion with nine parts of water. The above formula gives three gallons of emulsion, and makes, when diluted, thirty gallons of wash

"Another frequent cause of failure," continues Riley, "is the attempt to form an emulsion by churning together a small quantity of kerosene and a large quantity of diluent. Only a very unstable compound is thus formed. The very essence of the process requires that the oil shall be broken down by driving into union with it, a smaller, or at most an equal, quantity of the emulsifying solution, after which, if a genuine emulsion is formed, it may be diluted to any extent with water."

Persons who are intending to use this remedy will do well to heed the instructions given above, and to carefully follow out the directions.

CHAPTER IX.

How to Collect and Mount Insects.*

The necessary outfit of the Entomologist is neither a very large nor a very expensive one. It consists of a net, a number of bottles of alcohol, a few small boxes, one or two cyanide bottles, some pins and cork, and a few setting boards.

For a net, it is best to go to a tinner, and have him make you a wire hoop about twelve inches in diameter, with a socket, into which the handle may be introduced. The bag portion should be made of cheese-cloth, or, still better, Swiss muslin. The entire cost of the net will not exceed twenty-five cents.

If your specimens are to remain in the alcohol for any length of time, the purest, or 98 per cent., should be used, but if you intend leaving the insects in for a few days only, a poorer grade will suffice.

Next is your cyanide bottle, to be used in killing butterflies and such other insects as would be injured by immersion in alcohol. For this, get a wide-mouthed bottle or jar, and into it drop a few pieces of cyanide of potassium, (procurable at any drug store), and over them pour plaster of Paris, mixed with water. When this hardens it will form a smooth floor, the deadly fumes from which will soon overcome any insect placed upon it. The entire cost of this article will be about fifteen cents.

---

*Copyright, 1888. Noble M. Eberhart.

The other bottles and the boxes you can easily procure without cost.

The boxes are used to place insects in that it is desirable to keep separate from the others.

In collecting, sweep the fields with the net, and use it also for butterflies, and other insects on the wing. Many specimens lurk under stones and boards, and under the loose bark of trees and stumps. In the latter places many insects are found in winter, so that the collector need not cease his work at any period of the year.

At night many fine butterflies and moths may be obtained near electric lights, gas lamps, etc., and one method

Fig. 69.

resorted to is to smear a mixture of sour beer and molasses on the trunks of trees, and then go to these with a dark lantern, or a light hidden by wrapping a towel around it. When the place is reached, the lantern should be unveiled and the insects quickly collected.

After putting your Lepidoptera, (butterflies and moths), into the cyanide jar, do not put any beetles in with them, as the little feathers from the wings of the butterflies will injure your other specimens. As soon as dead, the butter-flies and moths should be taken out of the bottle and wrapped in little triangles made by taking a square piece of paper and folding it along one of its diagonals. The specimen is placed in the fold, the edges turned, to prevent it coming apart, and a number marked on it referring to the record.

The record book should be kept with great care, as on it depends the real value of the collector's work. In every bottle should be placed a slip, with a number on it referring to your record book.

One of the best ways of preparing the latter, is to rule off your pages and fill them out as the following example illustrates:

No.	Date:	Locality:	Remarks:
1	July 17	Electric light, Mayville.	Geopini, very plentiful.
2	" 18	" " "	A shower of Belostoma.
3	" 20	Lake near Jonestown.	
4	" 24	Woods near Aleta.	Found on carcass of a dead horse.

Now, having collected your specimens, the next step is to mount them.

For pins, get those made by Herman Klæger. No.'s 3 and 5 will do for all ordinary specimens, but a larger assortment of sizes is, of course, desirable.

In mounting beetles, the pin should be thrust through the middle of the right elytron, (wing cover), so that it will come out on the under side between the second and third pair of legs. After this, the limbs and antennæ should be placed in as natural a position as possible. (See Fig. 69.)

Now make a gauge out of a little piece of wood with a hole in one end about a quarter of an inch deep. This is to keep the insect at a uniform distance from the head of the pin, which is done by thrusting the head of the pin into the hole in the gauge, and then pushing the insect up to it.

On the pin beneath the insect should be placed a little piece of card bearing the number referring to your record book. In pinning Hemiptera, (bugs, cicadas, etc.), the pin should pass through the scutellum, or little triangular piece on the back. (See Fig. 70.)

Small insects, too small to pin, may be glued on the narrow end of a small piece of triangular card board, and a pin thrust through the wide end of the card.

Before pinning Lepidoptera, (unless recently collected), they should be softened, or rather, the joints should be re-

Fig. 70.

lieved of their stiffness, by a sand-bath. This is done by filling a little box half full of wet sand, with a thin paper over it. The specimens are laid on the paper, and moist air in a few hours renders the joints pliable. The pin is run through the middle of the thorax.

Lepidoptera are dried on a setting board, constructed of two smooth, flat pieces of board, a foot or more in length, laid side by side, with a narrow groove between them. On each end nail a strip, as seen in Fig. 71, and over the groove, on the same side that the strips are nailed, tack a strip of cork. Your setting board is now complete and your specimen should be placed with the abdomen in the groove, the

pin going into the cork.   Then carefully take the forewings and push them forward until the posterior margins
are in a straight line, that is, are at right angles with the
body.   Push the back-wings up to almost meet the fore
ones, and then pin a strip of paper over the wings to hold
them in place (Fig. 71), and leave the specimens for a week
or more, when the papers may be removed, and the wings
will retain their position.

Fig. 71.

The wings of dragon-flies, and sometimes the right
wings of grasshoppers, are stretched in the same way.

To mount larvæ, carefully press out the intestines, then
introduce a blow-pipe, or straw, into the anus and blow up
the body to its natural size.   Let it dry, and it will retain
this form.   Large specimens may be stuffed with cotton.

If you have collected some specimens which you intend
to keep in alcohol, pour off that which they were collected
in, and after washing them in clear water, put them into the
pure alcohol and they will keep as long as desired.

CHAPTER X.

A List of the Insects in this Work, Arranged According to the Plants they Infest.

APPLE.

Apple Aphis.
Apple Curculio.
Apple-tree Borer.
Canker Worm.
Codling Moth.
Flat-Headed Borer.
Palmer Worm.
Peach Curculio.
Plum Curculio.
Tent Caterpillars.

ASPARAGUS.

Asparagus Beetle.

CABBAGE.

Cabbage Butterflies.
Cabbage Plusia.
Harlequin Cabbage Bug.
Tarnished Plant Bug.

CHERRY.

Palmer Worm.
Peach Curculio.
Plum Weevil

CORN.

Chinch Bug.
Corn, or Boll Worm.

Corn Root-Worm.
Locusts.
Stalk-Borer.
Wire Worms.

COTTON.

Cotton Boll or Corn Worm.

CUCUMBER.

Melon Worm.
Striped Cucumber Beetle.

CURRANT.

Imported Currant Borer.
Imported Currant Worm.
Native Currant Saw-Fly.

GOOSEBERRY.

Gooseberry Fruit Worm.
Native Currant Saw-Fly.

GRAIN.

Army-Worms.
Cut-Worms.
Hessian Fly.
Locusts.
Stalk-Borer.

GRAPE.

American Procris.

Grapevine Flea-Beetle.
Grape Phylloxera.
Green Grapevine Sphinx.

MAPLE.

Cottony Maple-Scale.

MELON.

Melon Worms.
Striped Cucumber Beetle.

ONION.

Black Onion Fly.
Imported Onion Fly.

PEA.

Pea Weevil.

PEACH.

Peach Curculio.
Peach-tree Borer.
Plum Curculio.

PEAR.

Peach Curculio.
Pear Slug.
Plum Curculio.

PLUM.

Grapevine Flea-Beetle.
Peach-tree Borer.

Plum Curculio.
Plum Gouger.
Polyphemus Moth, or American
Silk Worm.

POTATO.

Potato or Tomato Worm.
Potato Beetle.

RADISH.

Radish Fly.

SQUASH.

Squash Bugs.
Striped Cucumber Beetle.

STRAWBERRY.

Strawberry Root-Worms.
Strawberry Leaf-Roller.

TOMATO.

Tomato Worm.

GENERAL FEEDERS.

Army-Worms.
Cabbage Butterflies.
Cut-Worms.
Locusts.
Tarnished Plant Bug.
Tent Caterpillars.

CHAPTER XI.

A Key to the Orders of Insects.

A. Mouth adapted to biting, *i.e.*, having jaws and mandibles.

    a. Front or upper wings horny or leathery; back wings membranous.

        b. Upper wings hard and opaque, forming a covering or shield for the under ones, which are folded (first like a fan, and then doubled), under them.

                COLEOPTERA (Beetles).

        bb. Upper wings somewhat thickened to protect under wings, which are folded fanlike, but *not* doubled.

           ORTHOPTERA (Grasshoppers, etc.).

    aa. All four wings membranous and transparent.

        c. Wings many-veined; abdomen not provided with a sting, nor with an apparatus for depositing eggs (ovipositor).

           NEUROPTERA (Dragon-flies, etc.).

        cc. Wings few-veined; abdomen usually provided with a sting, or with an ovipositor.

           HYMENOPTERA (Bees, etc.).

AA. Mouth fitted for sucking.

    d. Wings four in number.

        e. Wings having feathery scales.

           LEPIDOPTERA (Butterflies, etc.).

        ee. Wings not scaled.

           HEMIPTERA (Bugs, Plant-lice, etc.).

    dd. Two membranous wings.

           DIPTERA (Flies, etc.).

The spring-tails and bristle-tails, small wingless insects, are often classed as an order by themselves, called THYSANOPTERA.

THE MURRAY
School of Entomology

In affiliation with the Chicago College of Science.

The only Entomological School in the World.

COURSE OF STUDY.—Structure and classification of Insects; Comparative Anatomy; Injurious and Beneficial Insects; Sensibility and Intelligence of Insects; Collecting and Mounting Specimens, etc.

DEGREE AND DIPLOMA.—Persons satisfactorily completing the course, will receive from the Chicago College of Science, a Diploma and the Degree of Bachelor of Entomology. (Ent. B.)

FEES.—The tuition fees, including diploma, are $10 for the course, payable in advance.

FOR ALL INFORMATION ADDRESS

NOBLE M. EBERHART, Chancellor.

308 to 316 Dearborn Street,

CHICAGO, ILL.

This Course may be taken by mail.

ENTOMOLOGICAL WORKS

By Noble M. Eberhart, Ph. D.

Outlines of Economic Entomology.—A text book for Schools and Colleges, and a reference book for Farmers and Gardeners. Brief practical articles on the principal injurious insects, with the latest methods for their destruction. With 71 engravings on wood, made expressly for this work. 12 mo, cloth, 75 cts.

Eberhart's Key to the Families of Insects.—Simple, concise and practical. The only Key published. Illustrated. 12 mo, paper, 25 cts.

Some Curious Insects.—Suited to the general reader. 16 mo., cloth, Illustrated, 35 cts.

How to Collect and Mount Insects.—Full directions. Illustrated. 16 mo, paper, 10 cts.

Eberhart's Comparative Entomology.—Brief comparative descriptions of the various insect families, give number of joints each in tarsi, and in labial and maxillary palpi; style of antennæ and wings; abdomen with appendages; peculiarities, etc. (*In preparation.*)

Eberhart's Entomological Works.—A complete glossary of the technical words and phrases used in Entomological writings. Fully illustrated. (*In preparation; ready soon.*)

A. FLANAGAN, Publisher,

185 Wabash Ave., CHICAGO, ILL.

··INDEX··

A